FISH AND FISHING
IN ANCIENT EGYPT

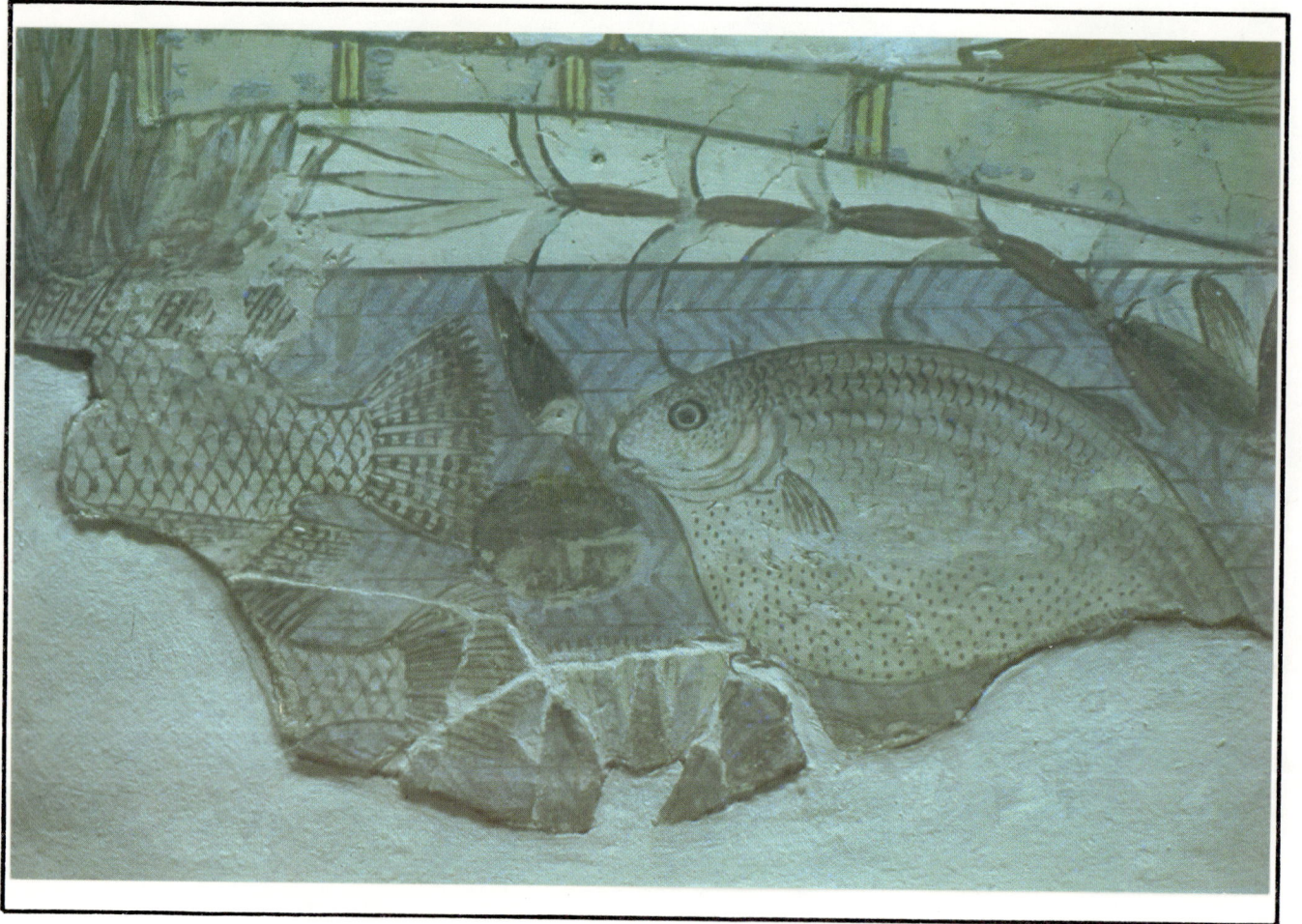

Tetraodon fahaka
Puffer fish

FISH and FISHING
in
Ancient Egypt

Douglas J. Brewer

Department of Anthropology & Museum of Natural History
University of Illinois

Renée F. Friedman

Near Eastern Studies Department
University of California, Berkeley

ARIS & PHILLIPS – WARMINSTER – ENGLAND

British Library Cataloguing in Publication Data
Brewer, Douglas J.
  Fish and Fishing in Ancient Egypt
  1. Egypt. Fish, ancient period
  I. Title II. Friedman, Renee F.
  597.0932

ISBNS   085668 399 X (cloth)
           085668 485 6 (limp)

Printed and published in England by Aris & Phillips Ltd, Teddington House, Warminster, Wiltshire.

CONTENTS

LIST OF PLATES

COLOUR PLATES

FIGURES

SOURCES OF FIGURES

Photographed by D. J. Brewer: I.1, I.2, I.3, I.4, 1.3, 2.41; by R. F. Friedman: 1.1, 3.11; by M.A. Hoffman: 1.2; by D. J. Brewer and E. C. Brock: 1.5, 1.6, 1.7,2.2, 2.3, 2.5ab, 2.6, 2.7, 2.8, 2.9, 2.10, 2.11, 2.12, 2.13, 2.18, 2.19, 2.20, 2.21, 2.22, 2.23, 2.24, 2.25, 2.26, 2.27ab, 2.29, 2.30, 2.31, 2.34, 2.35, 2.36, 2.37, 2.38, 2.39, 3.2, 3.3, 3.4, 3.5, 3.6, 3.7, 3.8, 3.9, 3.10, 3.12, 3.13, 3.14, 3.15, 3.16, 3.17, 3.18, 3.19, 3.20, 3.22, 3.23, 3.24, 3.25, 3.26, 3.27, 3.28, 3.29, 3.30, 3.31, 3.32, 3.33, 3.35, 3.36, 3.38, 3.41; Drawn by K. K. Brewer: 1.4, 1.8 after Davies 1901 pl XII, courtesy of the Egypt Exploration Society; 2.1, after Hobson 1987 (p. 42); 2.4 after Bates 1917, pl. X; 2.14 after Newberry 1893, pl. XXIX, courtesy of the Egypt Exploration Society; 2.15 after Radcliff 1926 (leaf) and Wilkinson 1883, no. 341; 2.16 after Winkler 1938, pl. XXXII; 2.17 after Boulenger 1907, fig. 26a; 2.32 after Vandier 1969, fig. 252; 2.33 after Davies 1901, pl. XII, courtesy of the Egypt Exploration Society; 2.40 after Davies 1902, pl. V, courtesy of the Egypt Exploration Society; 2.42 after Griffith and Newberry n.d., pl. XVI, courtesy of the Egypt Exploration Society; 3.1 after Greenwood 1966, fig. 1; 3.21 after Gaillard 1923, fig. 35; Photographed by E. C. Brock: 3.34, 3.37, 3.39, 3.40. 2.28 is taken from Petrie 1890, pl. XI.

ACKNOWLEDGEMENTS

We wish to express our thanks to the American Philosophical Society for the financial support of this project and to the Egyptian Antiquities Organization for permission to photograph the many tomb scenes required for this volume. We are also indebted to the Canadian Institute in Egypt and its Cairo Director Mr. Edwin C. Brock for logistical support and photographic assistance. The British Museum, London, the Egyptian Museum, Cairo, the Lowie Museum, Berkeley and the Egyptian Exploration Society provided the subjects for several illustrations. Ann Roth, Barbara Adams, May Trad, Bodil Mortensen and Michael A. Hoffman also supplied us with special photographs and pertinent information. Mindy James read several earlier versions of the manuscript and provided important editorial comments. Kristi K. Brewer deserves special thanks for donating her time, patience, and artistic skills to the many illustrations of this volume. Finally, we would like to express our sincere thanks to Adrian Phillips of Aris and Phillips for including this study in the Natural History Series and for the moral support that saw us through several hot summer field seasons.

INTRODUCTION

The physical environment of Egypt, hemmed in by desert and lacking extensive land for pasture and cultivation, undoubtedly contributed to the development of an efficient fishing industry. The Nile, the most important natural feature in Egypt, is the source of both its productive fisheries and agricultural fertility. To appreciate fully the development of fishing and how the Egyptians viewed and exploited fish, it is necessary to examine the unique physical conditions of the Nile Valley.

Before the Aswan Dam was built, the Nile was a seasonally inundated river plain. The rich, fertile soil that is characteristic of the Nile basin was deposited as a result of the annual floods. The accumulation of sediments (silt and clay) created a flood plain that, if viewed in cross section, would appear to be slightly convex.

Convex flood plains, such as the Nile, are marked by natural levees that constitute the low-water channel banks. These levees rise a few meters above the seasonally inundated alluvial flats. The lowest areas of the plain are more distant from the river and are often situated near the outer margins of the valley. These physical features were effectively used in the development of a basin irrigation system near the end of the Predynastic period.[1]

In mid-July the beginning of the annual flood could be detected at Aswan. The Nile would ideally rise and crest in southern Egypt by mid-August. Water would then spread out through major and minor overflow channels and through breeches across low levees, eventually spilling into the flood basins. The flood waters would reach the last basins at the northernmost end of the valley four to six weeks later. By early October, the first basins in southern Egypt were normally dry; by late November all but the lowest basin hollows in the northernmost parts of the valley were drained.[2] These four months of inundation were the first season of the Egyptian year. This was followed by the seasons of sowing and harvest named after the events of the agricultural cycle.

The physical characteristics of the Nile Valley also provided a seasonal habitat for fish, which in shallow flood waters were easily accessible to man. In Egypt's historic period, the retreating flood waters were equated with the creation of the world. Fish left stranded in shallow water pools as the Nile receded were associated with fertility and abundance, eternal life, and at times, primeval chaos.[3]

Nature of the Record

To trace the history and development of fishing and the use of fish in ancient Egypt, a variety of sources must be examined. Skeletal remains of fish are the main source of information during Egypt's prehistoric period and attest to the importance of fish as a means of subsistence. The analysis of these remains along with the limited number of fishing implements that have been preserved, aid in the reconstruction of Egypt's early fishing industry and culinary habits. Interpretation of this data is not, however, without problems. Differential survival of the delicate bones of certain fish can potentially bias the archaeological record, leading to inaccurate assessments of the types of fish captured and their relative abundance. For example, the high percentage of cranial remains compared to postcranial fish skeletal material recovered from many sites has led to speculation, possibly incorrectly, about fish preparation and storage practices of early Nile dwellers.

Fish bones as a group are difficult to identify to a given taxonomic level, and the vertebrae are among the most difficult elements of the fish to identify. Vertebrae are also not as densely constructed as the cranial elements and are, therefore, more subject to damage by post-depositional processes. Preliminary laboratory tests in the decomposition of fish skeletons show that vertebrae are the first elements to sustain damage and make identification difficult. These factors working together can contribute to a bias in favor of cranial identifications over vertebrae.

A perfect example of such a bias involves the Nile catfish (*Clarias* spp.). Cranial roof elements of this fish can be identified to genus, given certain geographical restrictions, even when broken into tiny pieces. Consequently, the mechanical breakdown of one *Clarias* skull could produce literally hundreds of identifiable *Clarias* remains. The same degree of breakdown to a *Clarias* vertebra or even the entire vertebral column, which in this case consists of 65 to 67 separate vertebrae (*C. anguillaris* and *C. lazera*) would, however, limit most identifications simply to "fish remains" or "fish vertebrae".[4]

Taphonomic processes (the processes that affect the preservation of skeletal remains) can seldom be clearly identified in the archaeological record and are not mutually exclusive. Such factors as scavenger behavior, the mechanical breakdown of the skeleton by fluvial transport, soil chemistry, and numerous others, including man, influence which bones are found and how many are preserved.

Given these circumstances, the paucity of vertebrae noted by several authors may be more a function of natural taphonomic processes than of the cultural practices of early Egyptians. Consequently, the common belief that fish were dressed and decapitated at the catch site and consumed at a later time should be re-evaluated.[5] A viable solution to the problem would be rigorous testing of fish-bone survivability across a series of environmental conditions and over a period of several years.

Partly due to changing subsistence patterns of the ancient Egyptians and to the changing focus of archaeological research in this period, evidence from skeletal remains in Dynastic times decrease. Written documents that provide information in the form of delivery accounts, tax receipts, medical recipes, onomastica, and, in one case, an account of a vacation spent in the marshes fishing and fowling are relatively rare and often incomplete.[6] To augment the loss of skeletal data and the rarity of written accounts, artistic representations become the major source of information on dynastic fishing practices.

Certain caveats, however, apply to fishing scenes used as historic evidence. Rarely is the main focus of the reliefs specifically to record the use of fish in daily life. Although a wealth of information can be derived from the depictions of fish represented on the tomb walls, the purpose of the representations was mainly related to religious and magical beliefs. Skeletal remains recovered from the historic period make this point quite clear by showing differences between the types of fish consumed and those chosen for depiction.[7]

Ancient Egyptians, from the earliest times, were keen observers of nature. This is reflected in the precise depiction of the many characteristics needed to identify Nile fish. It is also reflected in the symbolic or magical powers attributed to certain fish by virtue of their observed biological behavior. For example, mullets, having travelled from the Mediterranean Sea to the first cataract, were honored at Elephantine as heralds of the flood and as messengers of the flood god Hapy.[8]

The mouth-brooding habits of certain species of the genus *Tilapia* were also observed and associated with creation by the creator god Atum, who took his seed into his mouth and spat out the world. This perceived autogenesis was by extension associated with fertility and rebirth in the next life and the *Tilapia* in this connection becomes a favored motif in New Kingdom tomb paintings, amulets, and other minor arts. Particularly attractive are the faience bowls decorated on the interior with *Tilapia* and a lotus

bud issuing from the mouth, symbolic of new beginnings (fig. I.1).[9]

The brilliant breeding colors of *Tilapia* led to its association with the sun. The *Tilapia* or "red fish" was believed to accompany the solar boat as a guardian on its journey through the night and eventually the *Tilapia* was viewed as a form of the god Horus, who kills the enemies of the sun.[10]

According to some authors, the Egyptians may have been able to distinguish the habits of *Tilapia zillii*, a nest brooder, from other *Tilapia*. Possible depictions of nest brooding, although rare, range from the Predynastic to the New Kingdom. The three fish gathered around the glutinous ball of eggs in figure I.2 are thought by some to represent *T. zillii*.[11] An equally plausible explanation, however, is that the fish are mouth-brooding *Tilapia*, taking up the eggs after they had been fertilized in the nest.

The Nile catfish, because it favors muddy waters, was believed to guide the solar boat through the dark river of the underworld at night. Catfish-headed demons are depicted in New Kingdom royal tombs and numerous sarcophagi assisting the god Aker haul the solar disk on its nocturnal course (fig. I.3).[12] Catfish whiskers also reminded the Egyptians of a cat; according to classical sources, the catfish became a holy manifestations of the cat-headed goddess Bastet.[13]

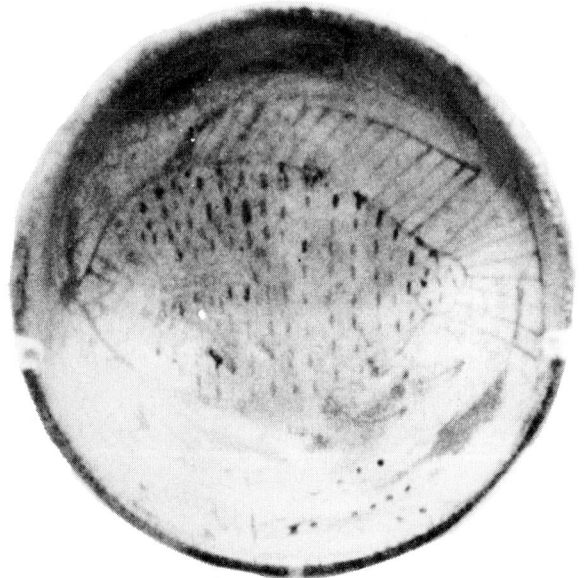

I.1 Tilapia and lotus bud – symbol of new beginnings. New Kingdom. British Museum, London.

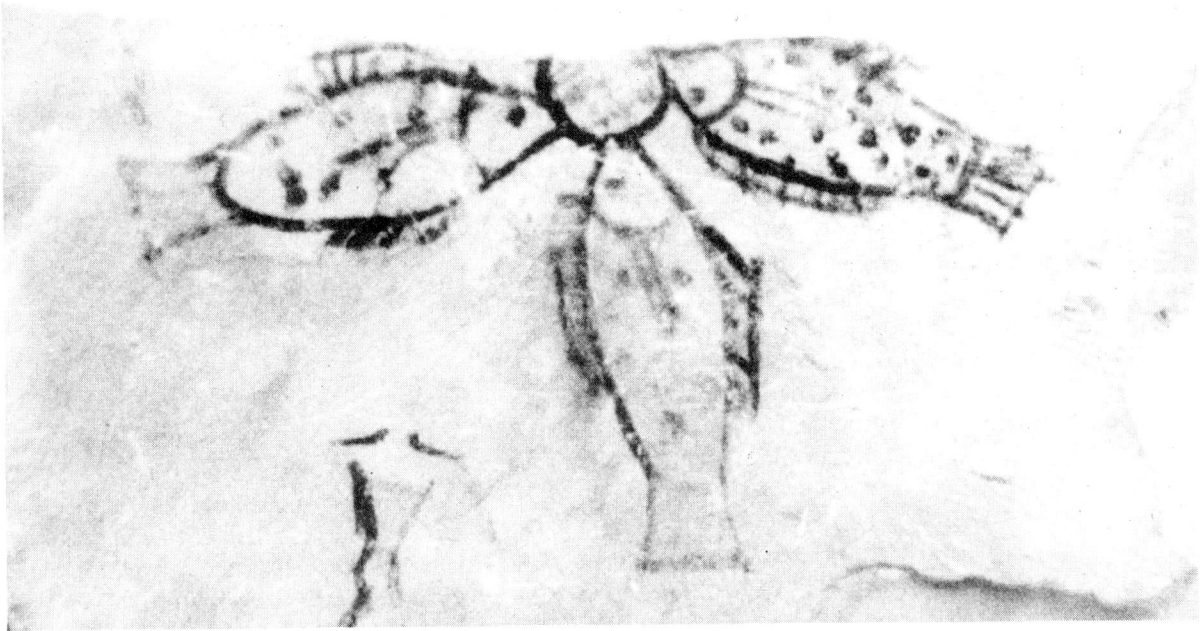

I.2 (top) *Tilapia and a mass of eggs(?) on an ostracon from Thebes. New Kingdom. British Museum, London. See Pl. I.*
I.3 (bottom) *Catfish headed demons depicted on a Late Period/Graeco-Roman period sarcophagus. Egyptian Museum, Cairo.*

The Egyptians were also aware of the peculiar habit of *Synodontis batensoda*, which swims upside-down, a characteristic depicted on several tomb scenes. In the Middle Kingdom, *Synodontis*-shaped ornaments, often made of precious metal, were popular. These were worn, usually by women, in the hair or as necklaces possibly as a charm to protect the wearer from drowning (fig. I.4). An amulet of this sort is mentioned in the ancient Egyptian tale about King Snefru and the magician. King Snefru (Dyn. IV) was bored and as a distraction went boating, using young maidens clad only in net dresses to work the oars. All went well until one of the maidens touched her braid and her turquoise fish pendant fell into the water. It was not until a magician was called in to roll back the water and retrieve the pendant that the boating party could continue its voyage.[14]

As the flood retreated, the abundance of fish caught in low-lying basins no doubt led to a connection of fish with fertility. Particularly those fish that preferred to live and breed in shallow waters (i.e., *Tilapia* and *Clarias*). At the same time, an association with the chaotic flood prior to the creation of the world (couched in the imagery of the flood receding and creation emerging), also placed fish in a sphere of chaos that needed to be controlled.

The religious conceptions about fish are, therefore, confusing and often contradictory. It is worth noting that apart from the goddess of the Mendes province, Hatmehit, who took the form of a *Schilbe* (cf. fig. 3.23), no other deities have a primary form represented by a fish. This is believed to reflect a partial or selective taboo on fish.[15]

In the Late Period, the association of fish with certain gods and goddesses in local myths is connected to the general popularity of animal veneration as manifestations of deities. The *Lates*, for example, was associated with the goddess Neith who at one point turned herself into a *Lates* to navigate the deep waters of the primeval ocean Nun. In her honor *Lates*, in mummified form, were offered as a token of worship, particularly at Esna where her cult was popular.[16] As a result, fish remains in the Late

I.4 Fish amulet or charm possibly used to protect the wearer against drowning. Note how the charm, when suspended, would orient the fish in an up-side down position. Dynasty XII. Egyptian Museum, Cairo.

Period again become an important source of information, but in a very different way than earlier. It appears that at the same places where certain fish were venerated and mummified, it was often forbidden to eat them.[17]

To what extent the religious view of fish influenced the use of this resource in daily life remains unclear. It is certain, however, that this type of data must be used with caution, for the beliefs attested in one time and place may not be valid in another. This is particularly important when dealing with the various classical authors (Herodotus, Plutarch, Aelian, etc.) who serve as the source of information after the scenes of daily life disappear from the tombs. It must be remembered that although their reports provide important information concerning the specifics of the fishing industry and fish consumption, the interpretations were made by "outsiders" and undoubtedly contain cultural biases.[18]

4

FISH IN ANCIENT EGYPT

Prehistoric Times

Archaeological investigation of Egypt's prehistoric economy has demonstrated that prehistoric Egyptians were well acquainted with their environment and made good use of the indigenous animals of the Nile Valley and desert for subsistence, as well as for raw materials in the construction of tools. Judging from the frequency of skeletal remains recovered from prehistoric sites, the prehistoric inhabitants of Egypt were successful terrestrial predators. Bones of mammals such as gazelle, wild cow, and hartebeest are common in many prehistoric sites. The pursuit of these animals was, however, not always the most energy or cost-efficient means of obtaining suitable food.

Cost efficiency refers to the amount of energy expended in the pursuit of subsistence versus the amount and quality of the food recovered. Although little archaeological evidence has been uncovered directly relating to the technological development of Paleolithic fishing, evidence based on the identification of fish bones recovered by archaeologists from Paleolithic sites suggests that certain taxa were part of an efficient and well-balanced seasonal fishing strategy, that would have provided a high yield of energy at a very low cost of time and man power.

A seasonal fishing strategy would most likely have been an integral part of a mobile settlement pattern whereby population or settlement shifts occurred in order to take advantage of resources as they became seasonally abundant. Ethnographic studies of modern hunting-gathering groups and many simple horticultural societies describe the practicality of this type of settlement pattern for maximizing cost efficiency. The expected archaeological evidence for such a pattern in prehistoric Egypt would include: an indication of a reoccupation of the same locality by the same social groups over a short interval of time; the occurrence of other localities occupied by these groups where different economic emphasis is indicated; evidence that various sites were utilized at different times of the year; and variations in the extent of the settlements, reflecting changes in the size of the social group during their annual cycle.

Using these criteria, a study conducted by Wendorf and Schild[1] in Wadi Kubbanyia (30km. north of Aswan) suggests that a seasonal occupation of sites had existed. Fish bones occurred at all seven Wadi Kubbaniya sites investigated, but they represent an important portion of the faunal remains at only one type of site: dune sites. The preponderance of fish at dune sites was explained by the physiographic setting.

Geological investigations have shown that the preserved slopes and front faces of the dunes were mantled with numerous thin veneers of silt separated by layers of sand and that the low areas between the dunes were covered by thick silt beds. These silts were deposited by the late summer flood, which at its maximum would partially submerge the dunes and fill the numerous concavities between the dunes and the wadi bottoms behind them. Fish were brought into these shallow pools with the flood water and became trapped as the flood receded. These fish could have been caught easily with spears, nets, or even by hand. Obviously, the dry summits of the dunes were an excellent place to camp while fish were being gathered.

This phase or seasonal occupation of dune areas, according to Wendorf and Schild[2], probably occurred in early September, just after the maximum flood. Fishing may also have been cost-effective at the beginning of the inundation, which corresponds to the breeding period of many fish.[3] Judging from the occurrence of fish remains in all of the Wadi Kubbanyia sites, fishing probably remained an important activity during all phases of occupation. However, the unique advantage of the dune traps provided by the flood led to an emphasis on fishing during this period.

The high frequency of mandibular (dentary and articular) and cranial elements in comparison to postcranial remains (vertebrae) suggested to the excavators that most of the fish were beheaded at the dune camp, were smoked or dried there, and were then eaten elsewhere. This phenomena, the delayed consumption of fish, was previously noted at Tushka in Nubia[4], a locality with very similar physiographic features to those of Wadi Kubbanyia. Additionally, fish remains at two sites near Esna, two dune localities of about the same age as the Kubbanyia sites but with a different lithic industry, suggesting different cultural affiliations, also contain a very high ratio of fish cranial bones in comparison to postcranial parts.[5] In light of this evidence, it was assumed that at late Paleolithic sites where fish were caught in large quantities, it was a common practice to remove the heads near the place of capture, to dry the bulk of the catch, to store it in a manner that is as yet unidentified, and to eat it elsewhere.[6]

Although this may be true in some cases, the little understood effects of taphonomic processes on these remains, render any such statements as speculative.

In all Paleolithic sites one family remains dominant, the Clariids (*Clarias* and *Heterobranchus*). This is, at least in part, a reflection of human selection, which in turn is closely related to a simple shallow-water fishing strategy. Today people collect these fish by hand, club, spear, and covered pots. Unfortunately, none of these implements have left or are the type to leave a distinctive mark in the archeological record.

Epi-Paleolithic and Neolithic Periods

Further insights concerning the role of fish in prehistoric Egypt has come from a recent study in the Fayum Depression, an oasis in the desert west of Cairo, dominated by Lake Qarun (Birket Qarun). The lake was connected to the Nile throughout most of its history and experienced annual changes in water level in concordance with the Nile.[7]

The first intensive archaeological investigations in the Fayum were conducted in 1926 by Caton-Thompson and Gardiner.[8] They concluded that two distinct cultures had occupied the area during the early to middle Holocene. However, recent research has shown that they placed the two cultures in the wrong chronological order. Fayum B, now known as the Qarunian is the earlier cultural complex (ca. 8220 +/- 105 bp to 7720 +/- 70 bp) and is characterized by the presence of backed blades, a lack of pottery, and no evidence of domestic plants or animals. Fayum A, now referred to as the Fayum Neolithic (ca. 6391 +/- 180 bp to 5070 +/- 110 bp)[9] is represented by a village-like economy composed of cereal grains and domesticated animals. The changes in material culture and the separation of Fayum A and B sites by a period of time devoid of cultural remains suggests that these assemblages reflect different peoples with no direct cultural relationships.[10] Yet they are similar to one another in one respect: the extensive reliance on fish for subsistence.

To understand the subsistence strategies employed by Fayum inhabitants, faunal remains were collected during recent archaeological investigations from a series of Fayum sites using systematic transect surveys running perpendicular to the former beach ridges of the ancient Fayum lakes. A total of five sites were incorporated in this analysis. Four sites were classified as Neolithic (Fayum A) and one site as Qarunian (Fayum B).[11]

The faunal assemblages, which were systematically collected from Qarunian and Neolithic sites, revealed that both groups relied heavily upon the lakes' fish resources and that they exploited the same types of fish in a similar order of ranked abundance.[12] This demonstrates similar food preferences or, more likely, similar subsistence strategies. It has also been suggested that the dependence on fish greatly affected the cultures of both groups. The uniqueness of the Qarunian lithics has been linked to its use in fish processing. The continued exploitation of fish in the Neolithic may also have been responsible for the relatively unimportant role cattle played in contrast to other contemporary sites.[13]

Based on the number of identified specimens per taxon (NISP), fish accounted for 94.4% (4171 NISP) of the entire Qarunian faunal assemblage and 71% (4585 NISP) of the Neolithic assemblage. Animals (mammals, birds, and fish) that prefer to inhabit shallow water environments accounted for 68.6% (7440 NISP) of the identified taxa. The Nile catfish (*Clarias* spp.) accounted for 97.6% (7264 NISP) of the shallow water fauna and 66% of the entire faunal assemblage.[14]

Clarias spp., the Nile catfish, prefers deoxygenated, shallow, swampy environments.[15] This type of habitat would have been present around a relatively large area of the prehistoric Fayum lakes when they were at a high level. The large quantity of *Clarias* remains identified suggests that Fayum inhabitants heavily utilized these shallow water-resource areas surrounding prehistoric Lake Qarun.

The Fayum fauna also suggest that both groups possessed a technology capable of obtaining fish from the deep, open water areas of Lake Qarun. This observation is based on the remains of the Nile perch, *Lates niloticus*. *Lates* generally inhabits deeper, more oxygenated waters than *Clarias*.[16] In fact, the oxygen requirements of *Lates* are such that it cannot tolerate the oxygen-depleted shallow water areas where *Clarias* would abound. Considering the habitat requirements of these two fish, the most effective way to exploit both taxa would be to use separate fishing methods: shallow water fishing for *Clarias*, and deep water fishing for *Lates*.[17]

Clarias, because they live in shallow water and are relatively large fish, can be effectively collected by spearing, netting, or by hand as a number of ethnographic reports describe.[18] *Lates*, on the other hand, which needs well oxygenated waters, would be collected most effectively by netting or angling in the open, deeper waters of Lake Qarun or the mid-channel areas of the Nile. Furthermore, *Lates* are known to exhibit a behavioral change once they reach a size of about 40cm. in length. Upon reaching this size class, they become very reclusive. They live in sheltered areas, such as rock crevices, and only dart out of these sanctuaries momentarily to strike at food.[19] Such behavior, in addition to the oxygen requirements of *Lates*, would most likely require a more sophisticated fishing technique than that used for the collection of *Clarias* to ensure a good catch.

Shallow-water and deep-water fishing strategies are further reflected in the faunal record by the recovery of a limited number of shallow-water taxa and a greater diversity of deep-water taxa. This suggests a selective strategy for the collection of *Clarias*, the predominant shallow-water taxon, and a more general fishing practice such as netting for deep-water fish.

Because *Clarias* remains dominated the shallow water fauna and because *Clarias* can be gathered year-round, an intensive study of the growth cycles of this genus was conducted to provide information as to the time of year it was collected. The seasonal study suggests that both Fayum groups collected *Clarias* at two different periods in the year: late spring-early summer and again in late summer or early autumn. Late summer/early autumn would correspond to the annual Nile flood and to the *Clarias* spawning period. *Clarias* would be highly aggregated at this time and after spawning, the fish would be sluggish and easy to catch. Late spring/early summer would correspond to low water levels, which would leave fish stranded in shallow pools left by a receding Nile.[20] A seasonal collection of *Clarias* during these two periods would offer the highest yield of fish per unit of expended human energy as well as requiring very little technology for the capture of the fish.

Seasonal settlement shifts have been suggested as an explanation for the different types of Neolithic Fayum sites that have been studied. Sites represented by one or two hearths located on or near the former shore of the lake are thought to have been seasonally occupied to take advantage of fishing opportunities.[21] Larger sites located further from the lake were thought to have been of a more permanent nature.[22]

Further excavations in the Fayum at a second series of five sites ranging from 6480 to 4820 B.C. found that in the earlier sites *Clarias* was the predominant taxon, followed by *Tilapia*. In the most recent site (4810 +/- 100 B.C.), however, there was a notable shift towards and increase in deep-water fish such as *Lates* [23], which may reflect a developing fisheries technology.

Neolithic sites in other areas suggest a similar reliance on fish, although detailed analyses on the fish bones are not as extensive. At el-Omari, fish were clearly an important protein source. Because of the great number of net sinkers recovered from the site, it appears the fish were gathered predominantly by nets. At Merimde an exceptionally large and diverse fish fauna was recovered.[24] In Upper Egypt, the Neolithic Badarian culture also relied at least in part on fish. Although the piscine fauna of Badari has not been subjected to analysis, *Lates* vertebrae and spines have been identified from the midden.[25] The spines appear to have been used as awls, a practice that continued into the Old Kingdom [26], and the vertebrae were modified and used as finger rings.[27]

The Predynastic Period

The later prehistory of Upper Egypt is separated into two units, Amratian (Nagada I) and Gerzean (Nagada II), which represent the beginning of an extensive cultural sequence leading into the historic period. These cultures are typified by permanent settlements, increasing social stratification, the beginning of irrigation technology, and an increased dependence on domestic plants and animals. Fish still constituted a part of the diet. The majority of the sites dating to this period are cemeteries, which were excavated in the early part of this century. The few settlements that were investigated produced fish remains, although no detailed analysis of the bones was ever conducted.[28]

Detailed information is available, however, from the extensive settlement remains at Hierakonpolis. Investigations reveal a strong dependence on domestic plants and animals over riverine and desert resources. *Lates* dominated the piscine fauna. Cranial and postcranial *Lates* remains were found 1.5km. from the river suggesting that the entire fish was transported to the site. A second excavation area produced numerous *Lates* vertebrae of significant size. These vertebrae came from fish that were certainly over 1m. in length. Large amounts of turtle and crocodile bone reinforce the impression that the occupants of this community, although heavily dependent on domestic food sources, were well adapted to the Nile and were able to exploit the large, and in the case of crocodile, dangerous river fauna.[29]

A unique look into the culinary habits of the Predynastic people is provided by human remains from the cemetery of Naga ed Der. The natural desiccation of bodies in Egypt's warm dry sand preserved not only the flesh of the interred individuals, but also the contents of their digestive tracts. Research on stomach contents showed that fish was occasionally part of these individuals' final meal. In one case, an eye, vertebrae, part of a fin, bones and scales of a *Tilapia* were identified, all apparently eaten in a barley soup. Another individual apparently consumed whole a large quantity of the tiny fish *Barilius niloticus* shortly before death.[30]

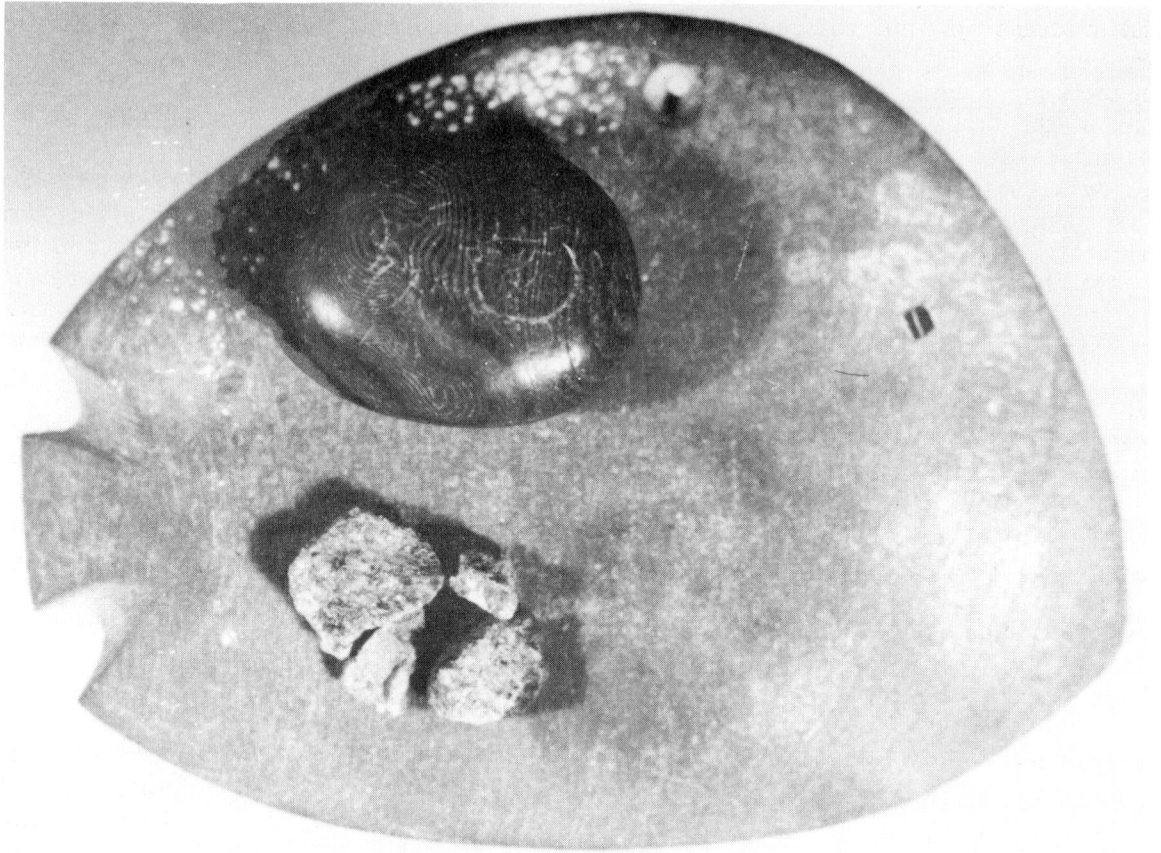

1.1 (top) Fish palette, grinding stone and green malachite from Naga ed Der. Predynastic. Lowie Museum, Berkeley.
1.2 (bottom) Hunted animal and fish on a ceramic sherd recovered from Hierakonpolis. Predynastic.

It is during the Predynastic period that material evidence for the consumption of fish declines and pictorial representations of fish become more prominent. The most common items modeled in the form of a fish were cosmetic palettes made from slate or graywacke. These palettes were used to grind malachite, a copper oxide of light green color, and occasionally red hematite (fig. 1.1). The resulting powder was mixed with fat or resin and used as eye makeup. Palettes are found in poor and rich tombs of both men and women. Often a small rubbing stone and sack of malachite accompany the palette. Cosmetic palettes are among the most frequent Predynastic grave goods. Originally they appeared in a rhomboidal form, but by Nagada II (Gerzean) animal shapes became popular. Fish-shaped palettes, however, were the most common and their popularity extended for a longer period of time than other shapes. The latest known fish-shaped palettes comes from Third Dynasty Nubia.[31]

Unfortunately, it is difficult in most cases to identify positively what taxon is being represented by the palettes. The general form, however, limits the range of possibilities to two common fish: *Tilapia* and *Lates*. Those palettes that are detailed enough to be identified almost always take the form of the *Tilapia*. In rare cases some palettes resemble the genus *Mormyrus* and possibly *Tetraodon*.[32]

Many different explanations have been offered for the predominance of fish-shaped palettes. Bates[33] noted that the palettes were mainly in the form of creatures that were hunted for food. Thus, he believed the palettes functioned as hunting amulets. The hunter would apply cosmetics ground on the palette to his body to insure a successful hunting or fishing trip. The decreasing importance of fish in the Gerzean period, however, makes this suggestion unlikely. Furthermore, hunting magic usually took the form of depicting the hunt or the prey animal, pierced by arrows, on pot sherds and rock faces. A ceramic sherd from Hierakonpolis shows a fish in the company of a hunted quadruped (fig. 1.2). Fish represented on ceramic sherds or in rock art are, nevertheless, relatively rare.[34]

A more religious association regarding the pisciform palettes is presented by other authors. Gamer-Wallert believes the palettes were actually amulets to be worn by the owner, although their size makes this unlikely. She bases this interpretation on the fact that most identifiable palettes take the form of the *Tilapia*, and she believes that Predynastic Egyptians like their historic counterparts associated this fish with eternal life.[35] This view is supported by a further connection between one of the ancient Egyptian

names for *Tilapia* "*wadj*" and a verb meaning "to be green, fresh, and youthful". This concept of youthfulness was not, however, restricted for use in the next world. The same word was also used in reference to the green cosmetic material ground on the palette.[36]

The use of the fish shape is also apparent in cosmetic containers, which have a long history in Egypt (fig. 1.3). Small ceramic jars in the shape of fish are also known from the Predynastic period and were apparently used to hold perfumes and salves that were, presumably, only available or affordable in small quantities.[37] Here a connection with the *Tilapia* and fish in general is more direct. The *Tilapia* is mentioned in the medical papyri as an ingredient for the preparation of a salve. Other fish are mentioned in the context of cures for other illnesses or physical shortcomings. The blood of *Synodontis*, for instance, was apparently used in an eyelash treatment and the bile of the unidentified "*abdu*" fish was used to treat the eyes.[38]

Early Dynastic Period (Dynasties I and II)

In the Predynastic period the first identifiable representations of fish appeared as well as inferences involving fish in religious contexts. By the beginning of the Dynastic period, fish and the hieroglyphic writing system were intimately linked. Fish appear not only as pictures of the fish they represent but as "letters" with phonetic values based on their ancient Egyptian name. These could be used to write words of similar sound, but unrelated meaning. Six fish, a small number in comparison to other animal classes, are used in this way: *Tilapia*, *Barbus* cf. *bynni*, *Mugil*, *Mormyrus* cf. *kannume*, *Petrocephalus*, and *Tetraodon fahaka*.[39] On some of the earliest texts known, *Tilapia* appears in the word "*inw*", (offerings or taxes). The *Mugil* also appears frequently in Early Dynastic seals in the title "administrator".[40]

Egyptians were very fond of word play. That the fish used as phonemes retained an association with the words they spelled or sounded like is reflected in the *Tilapia*-shaped cosmetic palettes and the green comestic material ground on them. The ivory and faience *Tilapia*, found in the royal tombs of the First Dynasties and as temple offerings, may offer another example.[41] They may represent "*inw*" (general offerings), or as amulets stand for rebirth and fertility. It is also possible that they may in fact be food offerings, as the little carved steatite basket of fish from Hierakonpolis suggests.[42] These interpretations are not, incidentally, mutually exclusive.

The most celebrated fish hieroglyph is the catfish used to write the name of Narmer, one of the earliest kings of Egypt.[43] Many early kings

1.3 Fish-shaped cosmetic containers. Dynasty XVIII. Egyptian Museum, Cairo.

had animal names. It is, therefore, unlikely that the catfish was merely a phonetic symbol. The exact translation of his name and its significance is unclear[44]: "Smiting Catfish" or "Beloved of the Catfish", although a connection with fertility is likely.[45] A catfish also appears on a small wooden plaque found at Saqqara (fig. 1.4). Here, five people walk in procession toward the name of King Djer. From the associated scenes of human sacrifice this may be either a funeral or a Sed festival where the king is magically rejuvenated. The men in the scene carry a lighted brazier, a statue, a catfish, a bird, and a spear. Although, the significance of these items is not clear, a procession of the gods or the royal ancestors has been suggested.[46] The last three items could, however, be associated with royal spear fishing (proof of virility and the ability to master chaos), evidence for which is also preserved on a plaque of similar age.[47]

Unfortunately, faunal remains from this period are lacking, as little archaeological work has been done on Early Dynastic town sites. However, fish remains in the context of funerary and food offerings have been recovered from one Second Dynasty tomb at Saqqara. The unplundered tomb contained the remnants of a meal laid out on individual plates next to the burial pit. A cooked fish, cleaned and dressed with the head removed, was among the foods served.[48]

The appearance of fish in this context is in direct contradiction to the pictorial depictions of funerary food offerings, which rarely depict fish. Funerary offering lists from the Early Dynastic period until the end of the dynastic era also never mention fish. The presence of a fish in this particular funerary meal has been unconvincingly explained by the tomb owner's lack of teeth, lost through gum disease, and the ease that fish could be eaten under such circumstances.[49]

Old Kingdom

In the Old Kingdom, the lack of fish in offering scenes is even more surprising given the popularity of fishing scenes in tombs. These scenes take two basic forms: the deceased tomb owner in the marshes harpooning fish and fishermen hauling in a net within a tableau of daily life in the marshes. Subsidiary to both types of scenes are fishermen on boats angling with hook and line, working handnets, weirs, or using basket traps.

By Old Kingdom times, Egyptian fishing technology was at a point where fishermen could selectively fish for particular taxa. Determining what fish Egyptians found most desirable has proven to be quite difficult. From the depictions of fish in tomb scenes, approximately 23 different

1.4   Reproduction of the wooden plaque recovered from Saqqara. Dynasty I. Egyptian Museum, Cairo.

forms of fish have been identified.[50] These do not, however, represent all the fish that inhabit the Nile, nor do they indicate the complete range of fish known and used by the ancient Egyptians.[51]

Of course, not all tombs depict all the fish represented in Egyptian art. Although no extensive study has been conducted on the frequency that different fish appear in netting scenes, a preliminary survey conducted on Memphite tombs does provide some information. On the basis of the Saqqara sample, Mugils are by far the most frequent fish represented and are the only fish found in similar scenes on royal monuments.[52] They are followed by Tilapia, Clarias, Synodontis, and Mormyrids. Lates, Citharinus, Schilbe, and Tetraodon fahaka are less frequently represented.[53]

In general, the ancient artists show a great deal of knowledge about the form and details of fish and fishing techniques, but at the same time inaccurately portray the types of fish that would most likely have been caught using particular fishing methods. For example, victims of the tomb owner's spear most often are a large Lates and a Tilapia. Fish that prefer very different habitats, and in the case of large Lates would not be easily caught by spearing. Scenes depicting the hand net, a shallow water fishing technique, also show Lates, which generally inhabits deep open and well oxygenated waters[54], a habitat not effectively exploited by a hand held dip net.

It is also likely that not all fish depicted in the tomb scenes were eaten. The ancient fishermen probably discarded unprofitable species

11

from their hauls, and artists may have portrayed only those that appealed to them or were familiar. Today, all the fish depicted in the tomb scenes can be found in the local markets.[55] Interestingly, even *Tetraodon*, a poisonous fish, can be found in modern fish markets; this fish is presented as part of the catch in the tomb of Kagemni.[56]

Although fresh unprocessed fish were sold in the market (fig. 1.5), large quantities of fish were dressed and preserved. Fish were usually dressed with the aid of a broad splitting knife (fig. 1.6), whose form strongly resembles that of some of the Predynastic and later flint knives. Fish were sometimes cleaned as soon as they were caught, but most were dressed ashore.

Without a doubt, due to structural differences in Nile fish, different taxa were dressed in different ways, and a variety of tomb scenes demonstrate this.[57] Nevertheless, a generalized description of the process based on the Egyptian's artistic examples can be offered. Fish were grasped by the tail and laid, belly down, on a block or on a flat-sloped dressing board (fig. 1.6). They were then cut in a downward stroke(s) along the vertebral column, after which the viscera was removed. The fish were then laid flat or hung to dry, the head and vertebral column, as shown in Old Kingdom scenes, often left intact. Removal of the vertebral column and head became more common in the New Kingdom. In some cases the fish was opened from the belly instead of down the back.

Another clue to the actual frequency or desirability of certain fish taxa is provided by fish-preparation scenes. Again the most frequent taxon identified from these scenes is the mullet[58], which aside from being highly prized for its flavor, also contained roe.[59] The roe, which was considered a delicacy, could be eaten fresh or dried. According to reports of the last century, to produce the dried caviar or *batarakh*, *Mugil* ovaries were placed in brine for about half an hour and were then pressed between two planks before being sun-dried. Vandier interpreted figures appearing in some *Mugil*-gutting scenes as representative of *batarakh* preparation[60] (fig. 1.8). Such scenes are, curiously, rare in Upper Egypt. This may be a reflection of the fact that gravid *Mugil*s were not abundant in these waters (spawning took place in the sea). After the Old Kingdom, fish roe no longer appears in the gutting scenes.

Some fish served as nourishment for the the marsh workers. Dried fish hanging next to tools in the temporary shelters of the marsh dwellers were no doubt for immediate consumption.[61] Another segment of the catch was used as payment. A scene from the tomb of Kagemni shows fishermen first taking their catch to be recorded by scribes; a portion of the catch was then distributed to various officials. Fish remaining in the possession of the fishermen could then be traded for other commodities; as market scenes in tombs depict[62] (fig. 1.5). Certain relative values of fish can be gleened from these marketing scenes. Fish could be traded for a loaf of bread[63], one fresh mullet for a jar of beer (fig. 1.7), and an entire basket of dried *Mugil*s for one amulet. Clearly, on the basis of the tomb scenes, fish was a cheap protein source, and New Kingdom documents confirm this.[64]

The fish depicted in the tomb scenes were not meant, however, just to represent the activities of daily life. These activities were also carried out expressly for the benefit of the tomb owner. This is made clear by the depictions of the tomb owner, several times larger than his employees, watching over his estate, while his ever-present overseers supervise the work. The representation of food production magically ensured the supply of food for the deceased in the next world.[65] Nevertheless, the absence of fish on the offering tables and among the servants bringing provisions for the offering table, places the exact purpose of the fishing scenes in question.

Perhaps fish were not intended for the tomb owner, but for his courtiers. Often shown behind the tomb owner spearing fish, are a series of friends or retainers. In some cases the names of these individuals have been preserved. Often they carry titles related to some function regarding the tomb. It is, possibly, to these officials and close friends that fish and other commodities were given[66] as a form of eternal payment for their services.

Consequently, fish, even if not specifically eaten by the deceased were still directly beneficial to him as a source of payment to insure the upkeep of his tomb, tomb offerings, and funerary estates. That proper payment had been made for services rendered was a matter that the ancients were very careful to make clear in both pictures and words.[67]

That wages in fish were paid to the upper management as well as the lower ranks is only problematic when one seeks to explain the tomb owner's apparent avoidance of fish. It has been suggested that fish was not a particularly valued food source among the upper class cattle barons and for this reason does not appear on the table of offerings.[68] The several occasions in which fish are being presented to the tomb lord may represent only an inspection of the catch[69], but because other commodities common in offering scenes are presented in the same scenes and in the same manner, it is difficult to substantiate this claim.

Perhaps the simplest explanation for the tomb

1.5 (top) Unprocessed fish sold in the market. Unas causeway, Saqqara, Dynasty V.
1.6 (bottom) Fish preparation in the Old Kingdom. Tomb of Two Brothers, Saqqara, Dynasty V.

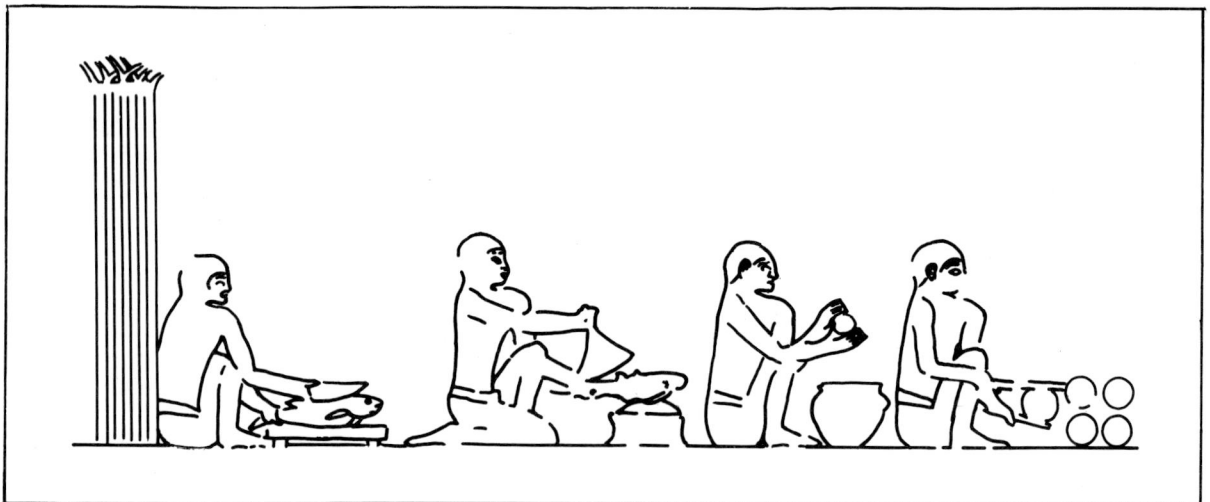

1.7 (top) Fish being traded for a jar of beer. Tomb of Two Brothers, Saqqara, Dynasty V.
1.8 (bottom) Preparing batarakh (?). Tomb of Urarna, Sheik Said, Dynasty V.

14

owners' avoidance of fish is that fish smell. "Behold my name (reputation) stinks, more than a catch of fish on a hot day." bewails a man in one text.[70] The ancient words for bad smells included a hieroglyph of a fish in the spelling as the determinative.[71]

The ancient Egyptians also used a fish in the writing of the word "*bwt*" which means forbidden, religiously impure, or taboo. It has been questioned whether the use of a fish symbol in the word "*bwt*", which is also the name of *Barbus* cf. *bynni*, was due to phonetic reasons or because fish were taboo. Authors have often argued for the latter on the basis of the exclusion of fish hieroglyphs in the Pyramid texts[72] (religious texts that decorated the royal tombs of the Old Kingdom). Other animals, however, are also excluded, perhaps to prevent them from magically coming to life and harming the funerary offerings. Restrictions on the consumption of "*bwt*" or animals holy to certain deities are only found in the Late Period and in passages written by classical authors.[73] Yet the absence of fish on the offering tables of the Old Kingdom has been considered an earlier manifestation of these beliefs.

From the sun temple of King Niussere (Dyn. V) and Middle and New Kingdom blocks from the Satet temple at Elephantine, we know that the *fahaka* (*Tetraodon fahaka*) and perhaps two species of mullets were venerated at Elephantine.[74] The absence of these fish in the local faunal record suggests an early consumption taboo, but other explanations are possible.[75] The sign for the Mendes nome, the *Schilbe* also suggests a fish cult for this area. Unfortunately, the faunal remains from these excavations have yet to be analyzed.

Perhaps due to a local taboo, or in the interests of ritual purity, Old Kingdom nobles or artists deemed fish inappropriate food in death. This is restricted, however, to the Memphite area, for the provincial nobles felt otherwise. Fish were part of their funerary meals and the fish catch was dedicated to their spiritual nourishment. A man bearing a fish in a tomb at Meir calls out "Make way for the fish for the spirit of the prince".[76]

Middle Kingdom

The restrictions regarding fish felt by the Memphite nobles disappear in the context of Middle Kingdom tombs. The Middle Kingdom Coffin Texts make it clear that the deceased did not avoid fish in the next life. The traditional offering table is, however, not altered to reflect this change, but the catch is regularly dedicated to the spirit of the deceased. It is also considered a gift of the marsh goddesses Sekhet and Hathor.[77] According to the Coffin Texts, the deceased wished to be like the crocodile god Sobek, who

lives on fish.[78] In another Middle Kingdom account, King Merikare's father tells his son that the gods made fish for man to eat and the faunal remains from Middle Kingdom levels at Elephantine and the Fayum substantiate the continued use of fish as food in this period.[79] Fish ponds were maintained in the Delta to supply Middle Kingdom nobles with highly prized mullets, and large sailing ships were employed to transport mullets to Upper Egyptian nobles.[80]

Even the king did not avoid fish and enjoyed fishing. A fragmented XIIth Dynasty text describes a royal fishing trip. The beginning of the text is missing, but apparently the courtiers urge the overworked king (probably Amenemhet II) to take a fishing vacation, and they bring him a man knowledgeable of the Fayum marshes. Finally convinced, the king mounts a fishing expedition taking all the princes and princesses with him.[81] Further evidence for lifting any restrictions on fish is offered by King Amenemhet III (?), who, perhaps in honor of some god of the Fayum, had himself sculpted in the guise of an ancient deity bearing an offering table overflowing with mullets.[82] It is also known that Middle Kingdom kings commanded that immense quantities of fish be distributed to various temples.[83]

Fishing scenes remains popular in tombs of this period and also appear as three-dimensional models (e.g. fig. 2.42).[84] Fishing scenes are now clearly associated with the satisfaction of the tomb owners' nutritional needs. A scene in one tomb may also be related to satisfying needs of another type. In the tomb of Oukhotep at Meir[85] the fishermen are not men, but women in men's brief clothing. One is reminded of the story of King Snefru and his female rowers clothed only in net dresses. Perhaps here we see the usurpation of such royal prerogatives at least in death, if not in life.[86]

New Kingdom

Scenes of daily life showing men hauling in large seine nets continue to be common in the first part of the New Kingdom, but are later replaced by scenes with more religious significance. In these later scenes, the tomb owner is often shown spearing or angling for *Tilapia*, a symbol of rebirth. *Tilapia* with lotus blossoms, also symbolic of rebirth, issuing from their mouths appear as a common motif on the interior of faience bowls, which have been found only in funerary contexts (see fig. I.1). Apparently, during the harrowing journey to the netherworld, the deceased at one point changed into a fish. The possession of fish, either real or a representation, aided the deceased in this transformation.[87]

During the New Kingdom, fish consumption is depicted more frequently than in any prior period. After a hard day of inspecting the fields, the New Kingdom tomb owner, unlike his Old Kingdom predecessors is shown consuming fresh fish as part of a massive midday meal.[88] Fish are also served to guests (both living and deceased) at wakes, but still rarely appear on the funerary table of offerings.

Kings and gods enjoyed fish as well. On the reliefs celebrating the Sed festival of King Osorkon II, the gods bring fish to him as gifts.[89] Fishermen in the service of the Temple of Khnum were released from paying taxes by the king.[90] Ramesses III, over a period of 31 years, donated some 474,640 gutted, fresh and pickled fish for the festivals in honour of Amun at Thebes, and another 19,600 fish to smaller temples throughout Egypt.

A number of New Kingdom texts mention fish in the context of daily life. Fish were issued as rations to the troops of Seti I.[91] Ramesses II, relating how well provided his workmen were, mentions that they have their own fishermen to keep them supplied. Some of these workmen were employed to work on the tombs in the Valley of the Kings and lived in their own village at Deir el Medina in Thebes. Excavations at this isolated site have produced a number of ancient documents that support the king's claim.

According to the documents, four times a month the artisans of Deir el-Medina received great quantities of fish, which constituted their principal nourishment. Twenty fishermen were contracted to supply the workmen and the fish were distributed proportionally by rank to approximately 40 individuals. Although the supply was not constant, over one six-month period the documents show that one fisherman could supply 9,700 deben or 882 kilograms of fish. This translates to 147 kilograms of fish per month.[92] The sizes and number of fish were also recorded and these figures have helped in the identification of some taxa. Fish that were supplied to the workmen include, *Tilapia, Synodontis, Mormyrus,* and possibly *Alestes*.[93]

The royal court also ate fish, and artificial ponds near the palace were stocked with fish. The palace storehouses at Amarna are also shown as being well stocked with fish.[94] The Hammamat Stela informs us of "the 200 officers of the court fishermen", attendants for Ramesses IV, whose task principally consisted of securing "a plenty of fish" for the enormous entourage of the monarch.[95] This included the royal harem as well as guests for various official functions such as the Sed festival.[96] Fish remains within the kitchen of the priests' quarters at Karnak suggest that the priests or their servants ate *Clarias, Synodontis, Tilapia,* and

other common Nile fish.[97]

Other texts mention the transport of cargo boats containing large quantities of dried and fresh fish destined for redistribution and sale.[98] A tomb painting depicts one of these ships with rows of dried fish hanging from the riggings.[99] When mentioned by name, the most frequent cargo is the mullet. This is not surprising, for mullet, according to texts, was more desirable and expensive than other types of fish, making its transport profitable.[100] Fish were also used as payment in international trade. In the report of Wenamun, 35 baskets of dried fish were destined as partial payment for a shipment of Syrian cedar.[101]

Given all the evidence for fish consumption during the New Kingdom, it is ironic that restrictions against the consumption of fish first appear during this period as well. Spells 64 and 148 of the Book of the Dead and in a magic papyrus of the Ramesside period contain instructions that "This spell should only be read by a man whose feet are clean, who has not been with a woman, who has not eaten goat or fish." These restrictions were, apparently, necessary to remain ritually pure. On the calendar of lucky and unlucky days, it was forbidden to eat certain fish on holidays and other mythological events. For example, on the 22nd day of the first month of the season of inundation, no fish were to be eaten because on that day Re created fish by eating the gods. When they upset his stomach, he vomitted out their bodies as fish and their spirits as birds.[102]

Fish veneration, already known in the Old Kingdom, continued to be expressed during the New Kingdom in the form of fish burials. At Gurob, an entire area of the cemetery was devoted to fish. The pits in which the fish were placed were more carefully made than those of the surrounding oxen and goat burials. One pit was lined with mud brick stamped with the cartouche of Ramesses II. The interred fish, in almost every case, were packed with fine grass ash, which probably served as a preservative. Fish were placed on a thick layer of this ash and covered with the same material. In the case of large specimens, the mouth and the openings behind the gill cover were also packed with ash. On a few of the largest fish, a slit was made along the ventral surface of the body and the cavity was stuffed with ash.

Most of the fish identified were *Lates niloticus,* a fish associated with the goddess Neith. A few specimens of other common fish were found whose religious associations are not known: *Synodontis schall, Bagrus docmac,* and *Clarias lazera.* In no case were different taxa placed together in the same pit. A few specimens were wrapped in cloth. This practice is illustrated in a tomb at Deir el-Medina depicting the jackal-headed Anubis embalming a large unidentifiable "*abdu*" fish.[103]

The Later Periods

In the later periods, fish in the daily life of the Egyptians did not differ substantially from that of the New Kingdom. The sources of information, however, change. Greek and Roman travellers now provide the majority of information. Diodorus stated "... the Nile contains every variety of fish and in numbers beyond belief: for it supplies the native not only with abundant subsistence from the fish freshly caught, but also yields an unfailing multitude for salting." Twenty-two different kinds of fish came from the Fayum lake alone.[104] These fisheries, operated as a royal monopoly, brought into the treasuries a talent of silver a day at the end of the inundation and a third of that sum every day during the rest of the year.[105] Aside from the royal interests in this revenue-producing industry, the people of Egypt continued to depend on fish. Herodotus[106] said that every marsh dweller was in possession of a net. Some people in the Delta including the aged, lived exclusively on fish. Fish was also the first food a child ate after weaning.[107] Other authors even provide ratings of fish palatability, biased, of course, by their own tastes.[108]

Fish were known to have been purchased as gifts to feed sacred animals, but certain taxa were considered holy and not eaten.[109] The classical authors also speak of the avoidance and condemnation of fish by priests.[110] Herodotus[111] wrote "The priests indeed entirely abstain from all sorts of fish, therefore, on the ninth day of the first month when all the rest of the Egyptians are obliged by their religion to eat a fried fish before the doors of their houses, they only burn them not tasting them at all". This taboo was apparently associated with the damage fish had done to the god Osiris.

Consumption Taboos

Sparked by the amusing accounts of the classical authors, interest in the taboos surrounding fish consumption has been great. However, only in the Late Period can taboos against fish consumption be substantiated. The question that remains is whether these restrictions were new to the later periods or extend back to the Old Kingdom and perhaps have roots in the Predynastic.

Around 750 B.C. the Kushite King Py (ankhy) conquered Egypt. When the Egyptian princes came to pay their respects and vow their vassalage, all were forbidden to enter the palace "because they were uncircumsized and were eaters of fish, which is taboo in the palace. But Prince Namart entered the palace because he was clean and did not eat fish."[112]

This evidence from Py's victory stela is the first explicit native account of a general taboo on fish consumption. Some authors have viewed this taboo as one imposed by the Nubian Py upon Egypt. However, the fact that a Delta prince shared the same belief shows it to be a custom practiced at least by some individuals on the local level. Different localities had different customs. As Athanasius reports, "the fish venerated by some, was in other places eaten as food".[113]

Later temple inscriptions corroborate the existence of these local injunctions, and provide lists of what was "*bwt*" (taboo) in the different provinces of Egypt. These lists mention six fish: *Lates*, *Tilapia*, Catfish, *Mugil*, *Tetraodon fahaka* and one still unidentified fish, which may be the eel *Anguilla vulgaris*. In some places it was forbidden to eat any fish. The classical authors also provide information about fish considered holy or forbidden and a mythological explanation.[114]

Plutarch's report, the only complete version of the Osiris myth, may illustrate the reason for the exclusion of certain fish from the diet in certain areas. The myth recounts that in the time of the gods, Osiris ruled the earth. His brother Seth coveted his throne and conspired to murder him. Successful in his plan, Seth cut up the body of Osiris and scattered the pieces over the length and the breath of the Nile. Isis, the wife of Osiris, was eventually able to recover all the parts with one exception, the phallus, which had been devoured by three fish; the Lepidotus, Phagrus, and Oxyrhynchus. Plutarch states "it was these very fish that the Egyptians were most scrupulous in abstaining." With the irretrievable loss of this vital member Osiris, despite the great magical powers of his wife, could not fully be brought back to life, although she was able to create a replacement part and impregnate herself to produce a son, Horus. Horus eventually avenged his father's death. Once avenged, Osiris took the throne of the netherworld.[115]

The basis for this myth was the cyclical pattern of agriculture. This important story was of great antiquity in Egypt and had earlier local variants all with a similar theme, the loss of the phallus and its consumption by fish.

In the New Kingdom Tale of Two Brothers, Bata is accused of molesting his brother's wife. To prove his innocence he cuts off his phallus and throws it into the water where it is eaten by a catfish. He then goes into exile. The gods take pity on him and send him a wife who deserts him at the first opportunity. He avenges himself by first transforming into a bull, then a tree, and

17

finally finds rebirth as the son of his own wife.[116]

The earliest variant of this myth, the "shepherd's song" is preserved in the Old Kingdom tombs at Saqqara. While his flock trample in the freshly sown grain seeds, the shepherd sings, "My *bty* is in the water among the fishes. It speaks with the catfish. It greets the *Mormyrus*. Oh goddess of the west (land of the dead) where is my *bty*?" Central to the understanding of this story is the "*bty*". With reference to the later versions of the story the "*bty*" (or *Bata* of the New Kingdom tale) can be identified as the seed, or the life force within the phallus. The purpose of the prayer is to resurrect the dead god of agriculture, later known as Osiris, and cause the seeds to sprout.[117]

The status of the fish associated with these myths is, however, uncertain. The classical authors provide anecdotes in which fish are venerated because of their association with Osiris as well as disdained for devouring his phallus.

A great deal of energy has been devoted to the identification of the fish mentioned by Plutarch in the Osiris myth. Of the three fish mentioned, only the Oxyrhynchus can be identified with certainty. Its name, meaning "pointed-nose fish", identifies it easily with the distinctive *Mormyrus*. The appearance of the *Mormyrus* in the shepherd's song and the account of Plutarch attests to the enduring nature of these myths. However, in the Old Kingdom the *Mormyrus* holds no greater significance than any other fish. It can be seen caught in nets, being sold, gutted, and offered to the tomb owner as are other taxa. Although reports on the fauna recovered from many historic sites are lacking, *Mormyrus* has been recovered from Elephantine, Hierakonpolis, and was among the fish delivered to the workers of Deir el-Medina. In classical times it was venerated in the town Oxyrhynchus, which took its name from the fish.

Aelian[118] recorded the custom of fishermen in the Oxyrhynchus area, "whenever fish were netted, they search the nets in case this famous fish has fallen in without their noticing it. And they would rather catch nothing at all than have the largest catch that included this fish ...". They also did not use fishhooks as a precaution against accidentally hooking a *Mormyrus*. Evidence corroborating this report can be found in a document dated to the first century A.D. Fishermen of this time were apparently required to sign an oath swearing not to catch the image of the divine Oxyrhynchus or the Lepidotus. Not far away, a cemetery of mummified *Mormyrids* was discovered at el-Omari. Further evidence of the veneration of *Mormyrus* comes from all parts of Egypt in the form of bronze statues, associated with the godess Hathor.[119]

In a conflicting statement, Plutarch says the Oxyrhynchus was hated for the harm it had done to Osiris. However, bronzes of this fish at Abydos, the holy city of Osiris, imply just the opposite. The evidence suggests that the association of Osiris with other gods of agriculture is of great antiquity and the placement of the Oxyrhynchus in direct opposition was a foreign innovation.[120]

The Lepidotus is more difficult to identify. A clue to its identification comes from its name, which means "scaley" and suggests that this fish possesses distinctive scales. On the basis of this evidence, it is likely that the Lepidotus is the carp-like *Barbus*. *Barbus* not only has large scales but a cult apparently venerated this fish, as numerous Late Period bronzes of its likeness seem to indicate. The manner in which the Lepidotus is depicted – a smooth dorsal fin spine, without posterior serration and the humped nape – equate it specifically with *Barbus bynni*, the most common and widespread taxon belonging to this genus. Mummified examples of this fish placed in *Barbus*-shaped sarcophagi have also been recovered from Thebes. One wooden statue of a *Barbus* contained the scales of the fish wrapped in cloth and stained with natron. Bronze statuettes found at Abydos serve to verify Plutarch's statement that this fish was associated with Osiris. However, the cult center for the fish, the city of Lepidotopolis, was located very close to Abydos. Consequently, the association of the Lepidotus to Osiris, a connection not mentioned in early versions of the myth, may simply be due to the proximity of the two cult centers.[121]

The identity of the Phagrus is unclear. Most of the information comes from Aelian. The Phagrus was said to be worshipped at Aswan. It was called the Maeotes by the people of Elephantine and was revered there because its appearance heralded the annual rise of the Nile. This report, corroborated by Plutarch[122], has been equated by Edel[123] to the Old Kingdom reliefs of the sun temple of Niussere, which depict the migration of two types of *Mugils* to Elephantine. According to classical sources, the Phagrus was also venerated in the city of Phagroriopolis located in the eastern Delta, an area famous for the production of *batarakh* (mullet caviar). In this town, the Phagrus was protected as a manifestation of the sun god Re. It was also considered holy in Heliopolis.[124]

The Greek name for the fish, however, means "voracious", which caused it to be equated with the catfish that ate the phallus of Bata and speaks with the *bty* in the shepherd's song. Furthermore, because *Clarias* (the Nile catfish) spawning behavior is triggered by the rise of the Nile, it too could be considered an indicator of the annual flood.

On the basis of the Tale of the Ibis, Rawlinson[125] equated the Phagrus with an eel. Briefly, the story relates the role of the black Ibis, which destroys the winged serpents that cross from Arabia to Egypt in the spring. Nevertheless, because of the covert political message of this tale, created at a time when Egypt had suffered greatly at the hands of foreign invaders, it should not be used as a reliable source for the identification of the Phagrus.[126]

Clemens of Alexandria provides yet another clue. He speaks of the Phagrus as a voracious fish with blood-stained fins. If we can infer that "blood stained" pertains to the color of the fins, this characteristic as well as the reference to its signaling the rise of the Nile, limits the possible identification to the family Characidae, or tiger fish. The two genera common in the Nile are *Alestes* and *Hydrocynus*. Both become more numerous during the Nile flood and both have red fins. *Alestes* and *Hydrocynus* are rarely depicted in tomb scenes and appear only in scenes from Upper Egypt.[127]

Other fish associated with important cults, but not involved in the Osiris myth include the *Lates*, worshiped as a form of Neith in the town of Latopolis (Esna), where thousands of mummified specimens of this genus have been found. And *Tilapia*, which according to Ptolemaic Temple lists, was associated with Hathor at Dendara.[128]

During the Late Period fish were obviously being venerated in some parts of Egypt. In other parts of Egypt they were trampled and mangled by priests as part of a festival ritual. Texts speak of putting the knife to their throats as embodiments of the evil god Seth and his entourage. This is no doubt where the report of Plutarch condemning fish originated. Other animals were also cast in the role of the fiend, as were all sacrificial animals destined for the god's table.[129] Since Archaic times, spear fishing scenes illustrate the suppression of chaotic elements of nature and the maintenance of world order. Additionally, the same scene portrayed notions of fertility, renewal, and rejuvenation: all things that exist can be exhausted but also renewed.[130]

The veneration of deites as embodied in fish appear to have a long history. To the Egyptians, the gods could manifest themselves endlessly. They could take up residence in one specific animal or all members of that species and the connection could be for any amount of time. It is only in the Late Period, however, that this manifestation is permanent, possibly leading to taboos.[131] Whether or not such customs were part of the folk beliefs that had been suppressed by the official religion until the Late Period is hard to assess.[132] Abstention from fish and other foods in other cultures, both ancient and modern, are, of course, known and served to forge group identity.[133] Such customs could have existed in prehistoric times, but information is lacking. On the other hand, the incursion of foreigners into Egypt in the Late Period may also have encouraged the Eyptians to separate themselves from their conquerors by following similar taboos and at the same time identify the enemy with the food not eaten. The situation remains confusing and only further analysis of the actual faunal remains can clarify the issue.

Throughout Egyptian history, the fish of the Nile were an important part of the diet. Perhaps not always prized by the upper class, fish served as a means of payment, reward, and national revenue. At the same time, the connection of fish with the cyclical life-giving forces of the Nile became an image in the Egyptian conception of the world.

19

ANCIENT EGYPTIAN FISHING TACKLE

Ancient Egyptian fishing implements have been recovered from archaeological sites that date from prehistoric times to the Graeco-Roman period. Evidence from a variety of sources indicate that by 2400 B.C. (late Dyn. V) harpoons, nets, traps, and hooks were all in use. By 1950 B.C. (Dyn. XII) the rod was also employed.[1] Unfortunately, it is impossible at this time to define a datable sequence for these fishing implements based solely on their recovery and archaeological provenience.[2] Nevertheless, some inferences about the implements and their use can be drawn from evidence obtained through the analyses of objects recovered through archaeological investigations, and from the representations of fishing implements in Egyptian art.

Fishing Spears and Their Construction

On the basis of evidence obtained through archaeological investigations and from the study of fishing implements portrayed in Egyptian art, spears used in the pursuit of aquatic game can be divided into three types: spears with a single head or point hafted to the end of the shaft, spears with a socketted head (harpoons), and two-headed spears (bident). Spear points of bone have been found at most Predynastic sites in Egypt from all periods of the Predynastic. Harpoon points of bone, horn, and ivory are also known from most Predynastic sites in Upper Egypt. They have been found in both tomb and settlement contexts but have only been recovered with any great frequency at Nagada.[3] In Lower Egypt, harpoon points have been found in all but the lowest level at Merimde and el-Omari.[4]

Spears with a single-hafted point were either unilaterally or bilaterally barbed. Bidents were unilaterally barbed. Spears with socketted heads (harpoons) were designed so that the point would come free of the shaft after the target has been struck, and the quarry was retrieved by a line made fast to the head.[5] Harpoon heads recovered from Predynastic Egyptian sites usually possess from one to three barbs, all aligned along the same side (fig. 2.1). Single-barbed harpoons are frequently illustrated on Upper Egyptian predynastic pottery, but actual examples are quite rare.[6] Even rarer are examples of harpoons with two barbs on one side. Bilaterally barbed points are rare prior to the Middle Kingdom[7], but Petrie has recorded a bilaterally barbed harpoon point believed to date to the Predynastic era.[8] This example and others were, however, apparently

purchased and little information is available concerning their true provenience. Their true age, thus, remains in question.[9]

Petrie attempted to place spear and harpoon points within his sequence dating system, a relative chronology derived from the analysis of pottery from Predynastic graves. Petrie placed the nonmetallic points in Sequence Date (SD) 38-63. Although he believed single-barbed copper harpoons existed as early as nonmetalic ones, he thought that the nonmetalic harpoons underwent a steady simplification in form throughout time. He placed the three-barb form at SD 38-63 (Amration/Nagada I – late Gerzean/Nagada IId), the two-barbed point at SD 45-53 (early Gerzean/Nagada IIa – middle Gerzean/Nagada IIc) and the single barbed variety at SD 49-63 (Gerzean/Nagada IIb – late Gerzean/Nagada IId).[10] The range of the broad-bladed "hippo" harpoon of copper, Petrie believed, lay approximately between SD 54-61 (Gerzean/Nagada II c-d).[11] The accumulated evidence seems to indicate, however, that when dealing with these implements, only broad chronological divisions can be made. The bone, ivory, and horn harpoons, regardless of type, range from near the beginning of the Predynastic period to the advent of historic times, when they were imitated in copper.[12] The more effective copper harpoons then began to supersede the nonmetalic harpoon heads at the very end of

2.1 Bone harpoon points from the Predynastic period.

21

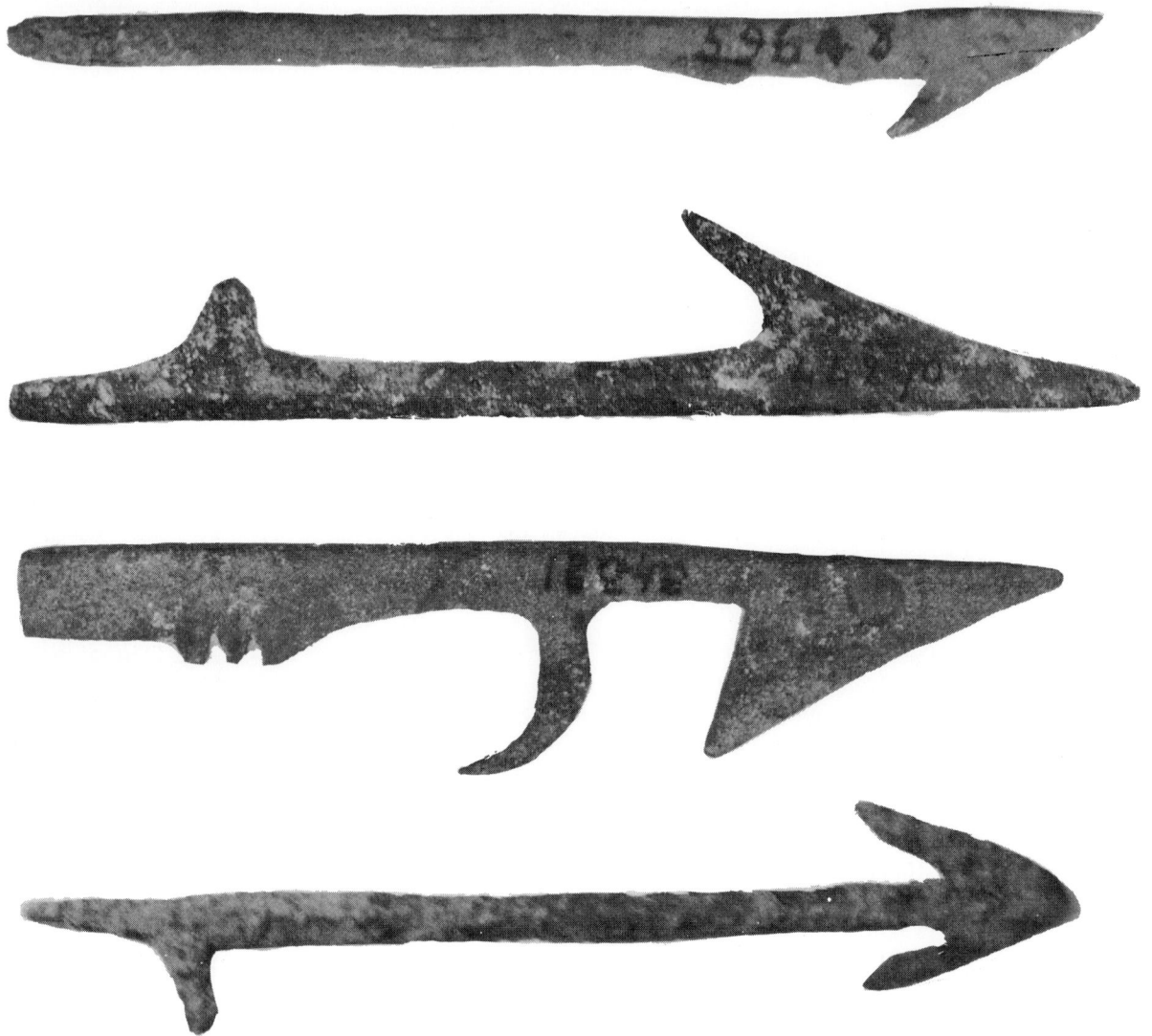

2.2 Four common forms of the copper harpoon. Egyptian Museum, Cairo.
(From top to bottom: 1.5x; 1x; 1.5x; .75x)

the Predynastic period (Nagada III)[13] (fig. 2.2).
Copper harpoon points were included among the
offerings in the royal graves of the Early Dynastic
period, and numerous non-functional models cut
from sheet copper were found in the tomb of
Khasekhemwy (Dyn. II).[14]

It is not possible to say whether harpoon
heads were used solely for fish or for crocodiles
and hippopotamus as well. The small size of
many of the early copper harpoons, the smallest
being only 6.5 to 7.5cm. long[15] suggests that they
were not used for large game. Although the only
large beasts in Egyptian waters were, in fact, the
crocodile and hippopotamus, very large Nile perch
(ca. 100kg. and larger) occurred in the river as

well. The smaller harpoon heads may have been
designed for use against these large fish; the larger
heads for the pursuit of the hippo and crocodile.
Unfortunately, conclusive evidence is lacking. It
would seem that separating hippo harpoons and
fish spears can best be accomplished on the basis
of hafting and their method of deployment, rather
than the shape and size of the head.

On the basis of tomb and temple scenes,
when hunting hippopotamus, the harpoon head was
attached to a retrieving line, and only loosely
fixed to the shaft. After the weapon was hurled
into the victim's body, the shaft became detached
from the head and the retrieving line was used to
track the animal. The line was either held by the

22

2.3 (top) Hippopotamus hunting. Tomb of Mereruka, Saqqara. Dynasty VI.
2.4 (right) The spade harpoon head.

hunter or attached to a float. Fish spears, on the other hand, were not hurled, but rather were jabbed into the water; the fisherman never released his hold on the weapon. A retrieving line is, therefore, seldom represented in the context of fish spearing and when present is restricted to provincial cemeteries in Upper Egypt. The most common spear represented in fishing scenes is the bident or two-headed spear.[16] In contrast, the harpoons shown in hippopotamus hunting scenes regularly possess a single head and barb (fig. 2.3), although in New Kingdom times a leaf or spade-shaped form can occasionally be seen (fig. 2.4).[17] These points, based on their representation in tomb scenes, were not used against fish.[18]

Symmetrical two-barbed harpoons are shown in scenes of the Middle Kingdom and New Kingdom, usually in association with hippo hunting, but occasionally in the context of spearing fish.[19] Actual specimens of bilaterally barbed harpoons have been recovered (fig. 2.2). By the Graeco-Roman period the harpoons depicted on temple walls were restricted solely to the spearing of hippopotamus and other animals associated with chaos.[20]

Through time, the stylistic and functional attributes of harpoons were reflected in their hieroglyphs. The single barbed harpoons with retrieving line, which was used to represent the phonetic value 'W^c' (i.e., wa), in the New Kingdom,

23

develops a bilateral barb corresponding to the form of actual contemporary harpoons. Another hieroglyph used as a determinative for the word 'bone' or 'harpoon' is depicted with two barbs on one side in the Old Kingdom, but by the Middle Kingdom was reduced to a single barb and a retrieving line was added.[21]

To date, of the spear and harpoon heads recovered from excavations, none appear to have been designed to be hafted in pairs. Thus, our knowledge of the bident or two-pronged fish spear is based wholly on ancient artistic representations.[22,23] Its absence from the archaeological record, however, does not necessarily imply that it is imaginery; few barbed points of any type have actually been recovered. If the ancient representations can be considered accurate, the use of the bident might well provide a clue to its rarity.

The first depiction of a bident appears on an ivory label dated to the First Dynasty. King Den is portrayed thrusting the bident into a pool of water.[24] The next extant representation of a bident, dated to Dyn. V, shows King Sahure on a boat with a mound of water, filled with fish rising in front of him. By presenting the water in this fashion, the undignified position of the king bending down to plunge the bident into the water can be avoided. And the king can maintain the traditional stance of smiting the enemy in this case, the chaos of marshes. This stance remained a characteristic symbol of royal might and power throughout the Dynastic period. That this motif was adopted whole-scale for the use of nobles in their tombs is clear from the inappropriate clothing worn by them, which includes heavy jewelry, the ceremonial false beard, and the royal shendyt kilt (fig. 2.5a).[25] In most cases the nobles also copied the covert political significant of the scene. Most examples of this scene depict a *Tilapia* speared on one point of a bident accompanied by a *Lates* speared on the other point. This is an interesting combination as one would not normally expect to find both taxa of such large proportions so near to one another as to be caught with the same thrust. Interestingly, the best fishing grounds for *Lates* would be the deep, well oxygenated waters of Upper Egypt. *Tilapia*, on the other hand, prefer shallow vegetated waters, indicative of the type of habitat that would be most abundant in the Delta. Thus, by showing a characteristic product of the Upper Nile, *Lates*, next to a product characteristic of the Lower Nile, *Tilapia*, a common theme in ancient Egyptian art was portrayed: the unity of two lands.

During the New Kingdom this political message was superseded by one reflecting a more religious theme. The scene itself remains quite similar with the exception that the *Lates* is replaced by a second *Tilapia*, which carries with it an association of fertility and rebirth.[26,27]

The frequent depiction of the tomb owner engaged in bident fishing does not, however, mean that the use of the bident was necessarily restricted to ritual purposes. The enduring popularity of the scene obviously had connections with actual recreational pursuits. Provincial tomb owners are depicted travelling to the Delta to take part in a sporting vacation. Carried in a sedan chair they are accompanied by retainers bearing throwing sticks and spear cases.

A scene from the tomb of Mereruka shows a small boat carrying three men holding bidents, each stocking a separate fish, which can be seen below the boat.[28] From this unique scene it is clear that the bident was occasionally used by the common man. Howecer only the upper class possessed enough leisure time to allow frequent use less cost efficient fishing methods. Therefore, the bident spear so often represented in Egyptian art may have been a comparatively rare fishing implement.

There appear to be two types of bidents represented in Egyptian art. In the first type, both points are equal in length and are lashed firmly to the end of the shaft with an elaborately designed series of cord wrappings situated between the shanks of the points (fig. 2.5ab).[29] The cord wrapping ensures that a fish caught in the spread of the tongs is held securely. The skill and care with which these spears were made is well illustrated by the XIIth Dynasty example from the tomb of Khnumhotep at Beni Hasan (fig. 2.5b). The twin heads of the bident each carry one barb. The points appear to be fitted into prepared slots on opposite sides of the shaft. Near the middle of the shank of each head, a lashing is attached; how it is attached cannot be determined. The lashings are in turn made fast to an object shaped in the form of a figure-eight, located between the spear heads. From the figure-eight, crossed cords are passed to the end of the shaft. The apparent purpose of this cording design was to keep the two heads of the spear from spreading too far apart when a strike was made.

In the second type of bident, one point is attached at the end of the shaft with a second point fastened some distance above the first point (fig. 2.6). The shank of the second point is apparently attached directly onto the spear shaft without an elaborate arrangement of cord wrappings. In some scenes this second point is not as long as the first point.

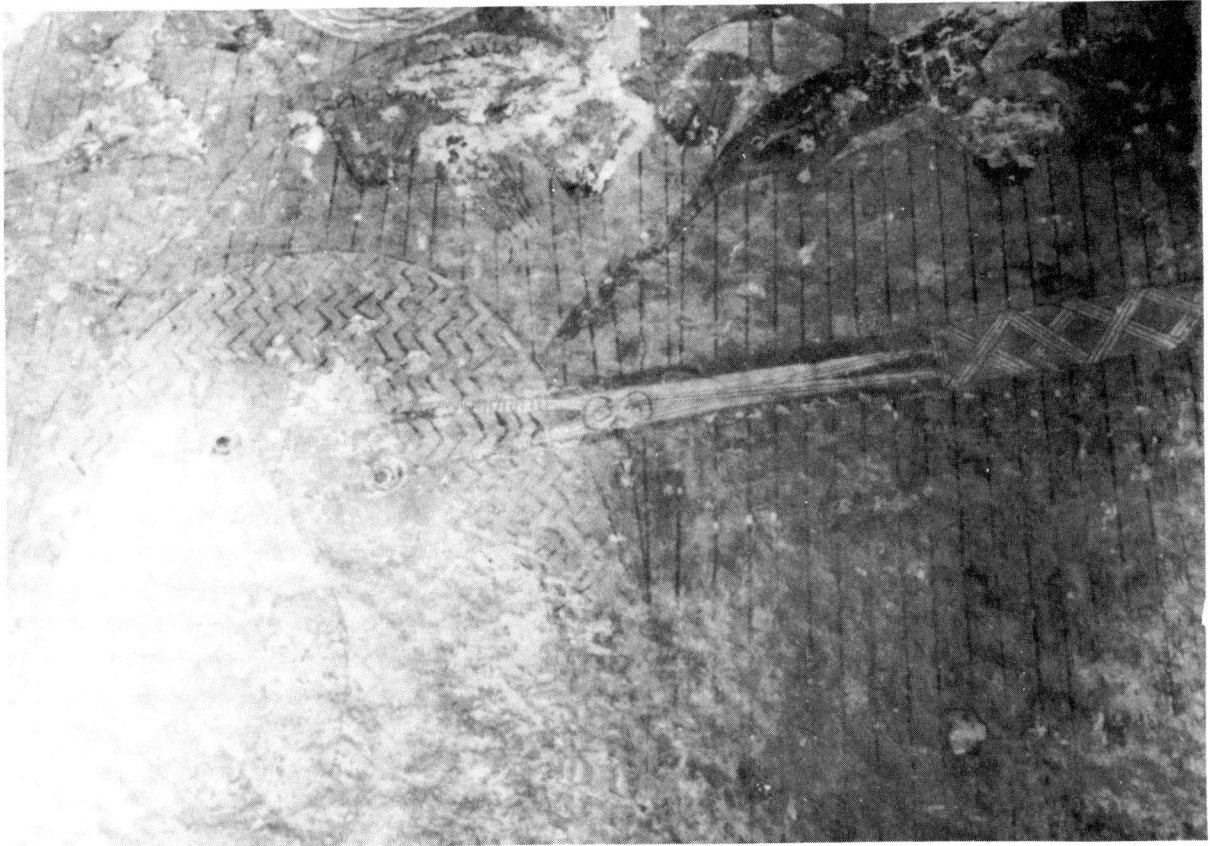

2.5ab *The bident spear. Tomb of Khnumhotep, Beni Hasan, Dynasty XII. See Pl. II.*

2.6 The second type of bident. Tomb of Mereruka, Saqqara, Dynasty VI.

The Hook

Fishing by hook and line is an ancient practice in Egypt. Hooks of various types have been found in Predynastic settlements from both Upper and Lower Egypt. Fishhooks of bone (fig. 2.7), ivory, and shell are particularly common in Predynastic sites in Middle Egypt.[30] Though absent from Fayum sequences[31,32], barbless varieties made from horn and bone, discovered at Merimde, indicate the early use of fishhooks in Lower Egypt.[33] In addition, a single fishhook of bone was excavated at el-Omari (fig. 2.7).[34] In earlier periods, gorges made of flint or bone may have been used.[35] Gorges are small spindles or double-pointed objects fastened at the end of a line that serves the same function, although not as efficiently, as the hook. Small double-pointed objects made of bone are not uncommon in predynastic sites of the Fayum, and some of these may have served as gorges.[36]

Near the close of the Predynastic period, the copper hook made its appearance; however, it remained relatively rare until Early Dynastic times.[37] The earliest hooks were of a simple shape, more angled than curved. The point was barbless and the head, formed on the same plane as the hook, was made by doubling over the end of the shank to form an eye. The hook's eye was generally open, although hooks with a fully closed eye are not uncommon. Examples of hooks from Old Kingdom Giza ranged in size from 1.5 to 2.5cm. across and 2.5 to 3.5cm. high (fig. 2.8).[38]

Hooks from the Old Kingdom generally lacked barbs. This style of hook lasted until Middle Kingdom times and was apparently still being used, although with much less frequency during the XVIIIth Dynasty.[39] The "gang" of hooks (combination of three or more hooks) depicted in the Tomb of Idout (Dyn. V) (fig. 2.9) is composed of barbless hooks, identical to those just described. The hooks in the hands of a vendor represented in another Old Kingdom scene

26

2.7 (top) Bone fishhooks from Middle Egypt and el-Omari (right). Prehistoric. British Museum, London and the Egyptian Museum, Cairo.

2.8 (bottom) The style of fishhook popular in the Old Kingdom. Egyptian Museum, Cairo.

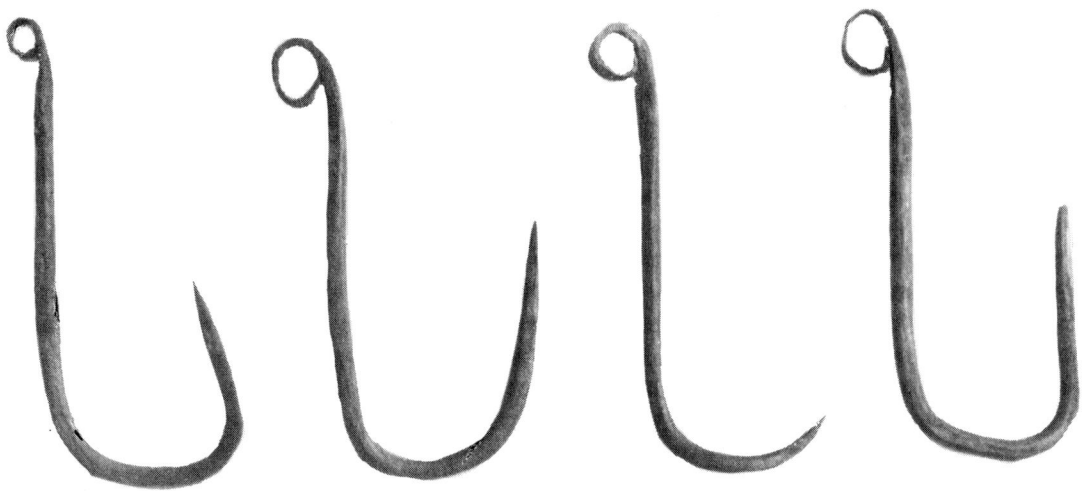

2.9 (top) Fishing with "gang" hooks. Tomb of Idout, Saqqara, Dynasty V.
2.10 (bottom) Fishhook vendor. Tomb of Two Brothers, Saqqara, Dynasty V.

from The Tomb of Two Brothers offers another example of the Old Kingdom barbless hook (fig. 2.10). Unbarbed hooks, as they occurred in the Middle Kingdom, are hardly distinguishable from those used in the Old Kingdom.

On the basis of archaeological evidence, barbed fishhooks became popular in Egypt around the XIIth Dynasty (fig. 2.11). The heads of these hooks are like the earlier ones, with eyes made by turning over the end of the shanks. The distal curve or bend is rounded or angular, and the barbs are well developed. With slightly modified proportions, these hooks lasted through the New Kingdom and even into the Late Period (Dyn. XXVI), although by this time they were often made of iron rather than copper.[40] New Kingdom hooks, however, often lack eyes. Instead, the end

28

of the shank was expanded slightly, presumably by hammering, forming a small flange with its greatest breadth in a plane at right angles to that of the hook (fig. 2.12). This design allowed the line to be fastened below the flange and when drawn up against it, the line could be held more securely than hooks of earlier designs, and would be less likely to wear through and break. However, Petrie noticed that some of the New Kingdom fishhooks that he excavated from Medinet Gurob, although possessing an improved design for attaching the line, were often not as well made as those of the Middle Kingdom.[41] By Graeco-Roman times the Egyptian hooks differed only slightly from those used by the classical world at large.[42]

Representations showing the use of the hook and line are not uncommon in tombs of the Old Kingdom, although such scenes are always subsidiary to the main scene.[43] Most often, the main scene portrays the tomb owner on a papyrus boat engaged in spear fishing or fowling. In an angling scene from the Old Kingdom Tomb of Ti, a fisherman is shown hauling in a large catfish (*Clarias* sp.). The man's left hand is grasping the line and his right holds a club, poised and ready to dispatch the fish once it has been landed (fig. 2.13). In an other Old Kingdom scene in the Tomb of Idout (fig. 2.9), the fisherman holds the

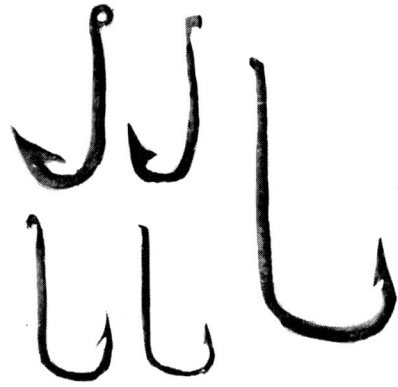

2.11 (top) Fishhooks of the type used during the Middle Kingdom. Egyptian Museum, Cairo.
2.12 (centre) New Kingdom fishhooks with expanded heads. Egyptian Museum, Cairo.
2.13 (bottom) Line fishing. Tomb of Ti, Saqqara, Dynasty V.

2.14 Pole fishing. Tomb of Khnumhotep, Beni Hasan, Dynasty XII.

2.15 A nobleman fishing. Tomb of Nebwenenef, Thebes, Dynasty XIX.

line with his index finger extended to feel the slightest tug on the line. The end of the line is armed with a gang of five hooks; on one a large *Synodontis* has been caught.

A survey of scenes shows that although several species of fish appear interested in taking the hook, representations of fish actually hooked are limited primarily to catfish, in particular the genus *Synodontis*.[44] Occasionally, however, the Nile catfish (*Clarias* spp.), *Tetraodon fahaka*, or in rare instances *Barbus* cf. *bynni* are shown caught by hook and line.[45,46]

In the Old Kingdom, fishing with hand lines seems to have been an activity carried out from water craft. The anglers themselves are generally represented, probably with some truth, as elderly peasants. Often they are shown with a receding hair line and wear a vest for warmth.[47] Angling with a hand-line was not restricted to the peasantry, but was apparently enjoyed by nobles during times of leisure. In the tomb of Mereruka an angler is sitting on a papyrus boat in the company of Mereruka's portly brother, who is clearly out for a pleasant cruise in the marshes. In the Middle Kingdom, hand-line scenes become less frequent, but examples show men fishing from the bank with a hand line, as well as from canoe. It is also in the Middle Kingdom that the rod is first depicted. The register from Beni Hasan (Tomb of Khnumhotep) shown in figure 2.14 is the earliest known fishing scene (ca. 1950 B.C.) depicting the use of a rod in Egypt.[48]

Examples of rod fishing continue into the New Kingdom, as evidence from Thebes indicates. Scenes from the New Kingdom Theban tombs, however, show the tomb owner himself engaged in rod fishing, rather than a commoner or servant. Additionally, fishing no longer occurs along the banks of the Nile or a canal, but within the confines of a garden; the line is placed into a well defined fish pond (fig. 2.15). In all New Kingdom scenes[49], the tomb owner is seated on a chair, often with his wife behind him helping with the rods or receiving the catch. Usually two rods with at least two lines were employed. Although all classes, most likely, engaged in fishing during their spare time, the New Kingdom scenes do not simply represent a pleasant outing or a means to gain sustenance for the deceased. Rather, they portray a message of rebirth made evident by the catch, which is always a *Tilapia*; the symbol of rebirth.[50]

If we can trust the scenes depicting the use of the fishing rod as representations of daily life, the fishing line was made fast to the end of the pole. It has been suggested that a running line or reel is being used by the man catching a catfish represented in the Tomb of Ti (fig. 2.13).[51] It is more likely, however, that the object in question

30

is a club. Large fish, according to the ancient fishing scenes, were often dispatched by club (cf. fig. 2.13).[52] The object in question and its counterparts in similar scenes appear to be shaped more like a baton than something that could be used as a reel. It is, of course, possible that the object could be used as both a club and a convenient place to wind the line.[53]

We have no information about what bait was used by the ancient Egyptian fishermen. Scenes that depict fish hooked by mouth suggest the use of bait. Wilkinson's[54] statement, "in all cases they adopted a ground bait, without any float" is not substantiated by the ancient scenes, where bait is never shown on the hooks.[55] A curious allusion to baiting, however, is mentioned in a passage in the Book of the Dead. The passage appears to read: "I have not caught fishes with bait of their own bodies".[56] Further evidence for the use of bait is found in a New Kingdom papyrus, which recounts an enjoyable fishing trip and mentions the use of bait. In this instance the method of fishing is by harpoon or arrow, suggesting that the bait was used to attract fish to the area.[57] Modern Egyptian fishermen dress their hooks with scraps of meat, minnows, shrimp, or with lumps of bread. In some cases they use no bait at all. Rather, they snag the fish by means of a naked hook attached to a hand line or trawl, or they "jig" the line; the fish is attracted to the shiny naked hook.

Fish traps

The use of traps to catch fish is a common practice in many parts of the world. A variety of fish traps were used by modern Egyptians at the turn of century and many records describe their use. Fish traps can be divided into two broad types. The first type of trap, a barricade trap, is designed to channel or barricade fish into a particular area or confined space. The fish are then collected by some other means, such as a spear or net. The second type of trap, the weir, is designed to capture and hold the fish. The fish are removed from the trap directly into a boat or container.

Barricade Traps

Examples of barricade traps from the ethnographic record indicate that they are generally restricted to aquatic areas that have a gradually sloping bed and a healthy supply of reeds or sticks. In ancient Egypt, these conditions were present in two large areas, the Fayum and certain portions of the Delta. The Fayum's environment from prehistoric times into the twentieth century was characterized by a lush cover of reeds and other forms of shallow water vegetation.[58] Similar conditions no doubt existed in the Delta.

The construction of simple barricade traps requires little technical skill. Ruffer[59] believed that spear fishing during the Predynastic and Dynastic periods was practiced on fish that had been channeled into shallow waters by means of barricades constructed of reeds and sticks. The fish, confined to a limited area, were then easily speared or captured by hand. Unfortunately, the durability of the materials used to construct a barricade trap and the areas where they would have been used provide us with little hope of ever being able to recover or perhaps even recognize such a device in an archaeological context.

Representations of barricade traps in Egyptian art are unknown. Winkler[60] suggested that two late Paleolithic petroglyphs, situated along the ancient migratory route connecting the Nile Valley with the Western Desert oasis of Khargah, might represent fish traps (fig. 2.16). Unfortunately, the accuracy of the identification cannot be ascertained. Even though pisciform drawings appear at the same site, their association with the "traps" is unclear. A possible piece of evidence, perhaps pertaining to the use of barricade traps, is provided by Strabo.[61] He states that the fish called Cestreus was captured during its spawning run by fenced enclosures.

This practice may have a modern parallel. Loat[62] recorded the capture of *Mugil cephalus* (grey mullet) during its spawning run by netting

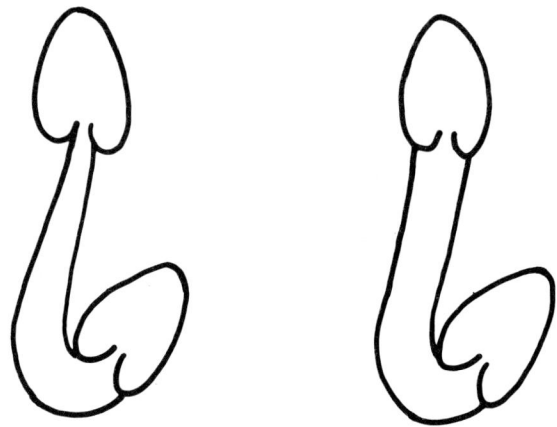

2.16 Winkler's barricade fishtrap.

the fish in river channels leading to the sea. After spotters noted the presence of a large school of mullets moving toward the sea to spawn, fishermen would set nets across the relatively narrow channels and encircle the fish. According to the fishermen, the process was aided by porpoises. The porpoises would wait on the seaward side of the lake's mouth and scare the fish back into the lake. Additionally, they aided in scaring fish that had jumped over the top of the net back into the enclosing circle of the net.[63]

It is difficult to say whether Strabo witnessed this event (or one similar to it), which would be considered a special case of "netting" fish, or if he indeed correctly identified a barricade trap of some type. In contrast to this example, Darby[64] described the use of a spiral trap used by modern Egyptian fishermen at Lake Mariut. These traps were set in shallow water and made of reeds arranged in a curved wall that formed an ever-tightening spiral. The fishermen would thrash the water from their boats to drive the fish toward the mouth of the spiral trap, where they were channeled into the center of the enclosure and then netted with ease.

The Weir

The wicker fish trap or weir has been employed throughout the world. Loat[65] reported at the turn of this century that weirs were used on

2.17 (top) The weir trap "garaby", used by Lake Qarun fishermen at the turn of the century.
2.18 (bottom) A representation of the small weir fishtrap from the Old Kingdom. Tomb of Two Brothers, Saqqara, Dynasty V.

the streams that entered into Lake Qarun. One called *garaby* was made of dried reed stalks lashed together in the shape of a cone (fig. 2.17). It had two openings arranged in such a way that when the fish passed through the smaller opening, it was unable to find its way out. At the narrow or closed end of the trap, the reeds were held together by a cord that could easily be undone to remove the catch. To set the trap, a barrier was usually built across the stream or channel, restricting the fish's passage to the opening of the weir. The trap was generally positioned to catch fish moving upstream.

Tomb paintings of the Old Kingdom picture the use of two types of fish traps.[66] The first type was small, of very simple construction and superficially, at least, resembles the weir described by Loat (fig. 2.18). The second type, which is less commonly represented than the first, was a trap of great size needing several men to work it (fig. 2.19). The weir is not a common subject in Egyptian art and is restricted primarily to the Old Kingdom.[67] The first known representation of a weir is in the Sun Temple of King Niussere (Dyn. V).[68]

Figure 2.18 illustrates the general shape and construction of the small ancient Egyptian weirs. At frequent intervals throughout their length, these traps were bound with rope; the small opening at the end was tied with a strong cord. This feature

2.19 The large weir trap of the Old Kingdom. Tomb of Two Brothers, Saqqara, Dynasty V.

2.20 *Tying the neck of a fishing weir. Tomb of Mereruka, Saqqara, Dynasty VI.*

is often omitted by the artist, but in figure 2.20 the man on the left working on one of the weirs is shown tying up the neck of the trap with a cord. This cord, attached near the rear of the weir around its narrowest part remains closed while the trap was being used. To remove the fish from the trap it was, of course, necessary to untie this cord. After untying the trap, its contents could be dropped through the bottom, or narrow end of the trap, into a basket (fig. 2.21).[69] The internal funnel of the ancient weirs, as illustrated in Loat's ethnographic example (cf. fig. 2.17) are not shown in the artistic representations, but some similar construction must have been present for the traps to have been effective. Small weir traps were probably weighted and set on the bottom of shallow water canals. They are not generally represented as being fitted with floats, but may have been propped up at their terminal end as shown in a scene in the Tomb of Two Brothers (fig. 2.18). The small wicker fish traps could apparently be set facing either upstream or down stream, the intended prey, most likely, dictating the appropriate direction. In figure 2.20 a weir is being set in a small canal, and from the representation of two traps facing right and two facing left, it might be inferred that the ancient Egyptian fishermen were attempting to catch fish swimming either up or down stream. In figure 2.22 the traps are all set facing the same direction.

34

2.21 (top) Emptying the catch from a small fishing weir. Tomb of Ti, Saqqara, Dynasty V.
2.22 (bottom) Three weirs set in position. Tomb of Two Brothers, Saqqara, Dynasty V. See Pl. III.

2.23 (top left) A large wier fishtrap fixed into position by stakes located on each side of the channel. Tomb of Two Brothers, Saqqara, Dynasty V.

2.24 (bottom) The tending of a large weir. Tomb of Ti, Saqqara, Dynasty V.

2.25 (top right) Removing fish from a large weir. Tomb of Two Brothers, Saqqara, Dynasty V.

The large weirs were of a more complex structure than those just described. From the representations it is difficult to determine exactly how they were built and manipulated. They were apparently fixed into position by a stake located on the bank at each side of the river or canal (fig. 2.23). The mouth of the trap (depicted in figure 2.24) was bound with a rope that seems to be tied off at the top of the opening. Near the middle of the weir was a large float, which helped support the large funnel-shaped trap. A fish, after entering the mouth of the trap and having passed into the funnel, would be caught and held in a secondary chamber located between the two roped neckings (fig. 2.24). In long weirs, the two-rope construction of the chamber would have been advantageous because it would facilitate removal of the fish by lifting the narrow end of the trap from the water (fig. 2.25). Details of the internal construction of the trap are never illustrated and it is not clear precisely what action is being taken by the two men pictured in figure 2.24. It is possible that the men are preparing to empty the contents of the trap by pulling on the rope, which would have sealed off the captured fish's only avenue of escape.

In order to pull up and empty the large weir, it appears, based on the actions depicted in the scenes and the conversations between the fishermen, that two to three boats were first maneuvered side by side. Artistic convention, however, depicts the boats in profile, facing each other. Keeping the boats together was a difficult task and an oarsman in the Tomb of Ti is encouraged by his companions to "row strongly ... make the boats come together well." As the boats are held together by the efforts of the oarsman, other men pull up the neck of the weir and empty its contents into awaiting baskets, as shown in the Tomb of Two Brothers (fig. 2.25).[70]

Basket Traps

One of the simpler means of obtaining fish, on the basis of artistic representations, was the basket trap. The basket trap was merely a basket of wicker design that in shallow water could be placed over a fish, entrapping it. The fish could then be grasped by the hand and removed from the trap (figs. 2.26abc, 2.27ab).

Only three examples of the basket trap in use are known from Egyptian art.[71] One example, an interesting fragment of an Old Kingdom scene now in the Cairo Museum, almost appears to be an instruction manual. The scene shows men standing knee deep in water, engaged in basket-trap fishing. To assist in comprehending the action taking place, the baskets are transparent, revealing an entrapped fish. In the first scene the man has trapped a *Tilapia* with the

2.26abc *The basket trap in use. Saqqara, Dynasty V. Egyptian Museum, Cairo. See Pls IV & V.*

2.27ab The basket trap, used in fishing. Tomb of Two Brothers, Saqqara, Dynasty V.

basket (fig. 2.26a); the second fishermen is shown pulling a *Clarias* out of his trap (fig. 2.26b). The third scene shows the fish being carried away (fig. 2.26c). Another scene from the Tomb of Two Brothers conveys a similar message, but the baskets are opaque (fig. 2.27ab).[72] Although the basket trap was rarely depicted and does not occur later than Dynasty XI, it was the subject of an aetiological spell in the Middle Kingdom Coffin Texts. In Spell 158, for some unknown reason, the goddess Isis has cut off the hands of her son Horus and has tossed them in the river. The crocodile god Sobek has then been asked to retrieve them with the use of a *ḥ3d*, a basket trap, and thus we are told how the basket trap came into existence. A celebration of this event took place on the 2nd and 15th day of every month, at which time it may have been customary to go fishing and fowling in honor of Sobek.[73]

Nets

Netting was undoubtedly a more cost-effective means of obtaining fish than either angling or spearing. Practically every kind of net known to the ancient world has at some time found employment in Egypt,[74] particularly in the calm waters of the Delta. The first archaeological example of a complete net comes from Neolithic el-Omari.[75] The net depicted in the tomb of Rahotep at Medum (fig. 2.28), dated to the early IVth Dynasty, is the first undisputed portrayal of a fishing net in Egypt.[76]

Actual specimens of net twine prepared from flax and other vegetable fibers were discovered at Kahun (Dyn. XII) in balls of two- and three-stranded string. Fragments of nets having 1.2 to 1.9cm. mesh have been recovered from the same locality (fig. 2.29). Even parts of nets having as fine as a 0.3cm. mesh were recovered. Kahun also yielded fragments of more recent nets (Dyn. XVIII) with meshes from 0.5 to 1.5 cm.[77]

Hand Nets

Hand nets were most likely used for taking small to medium-sized fish in shallow water or near the surface of deep water areas. Several varieties of open-mouthed nets are depicted, but the basic form was essentially the same during the Old and Middle Kingdom. In almost every case, the frame of the hand net consists of a pair of sticks crossed and lashed (fig. 2.30) near the handle, forming a "V". Between these sticks, which formed two sides of the frame, a third stick was employed as a brace. The projecting ends of the "V" were connected by a cord forming the third side of the triangular frame. A net was attached to this pre-formed triangle; the net could be opened or closed by manipulating a

Pl I fig I.2 (top) Tilapia. New Kingdom, British Museum, London.

Pl II fig 2.5a (centre) The bident spear. Tomb of Khumhotep, Beni Hasan, Dynasty XII.

Pl III fig 2.22 (bottom) Three weirs. Tomb of Two Brothers, Saqqara, Dynasty V

Pl IV   fig 2.26a (top) The basket trap with a Tilapia.   Saqqara, Dynasty V, Egyptian Museum, Cairo.

Pl V   fig 2.26b (centre) The basket trap with a Clarias.   Saqqara, Dynasty V, Egyptian Museum, Cairo.

Pl VI ÷ fig 2.35 (bottom) A seine net in which are (top to bottom l. to r.): Tilapia, Mugil, Synodontis, cf. Gnathonemus, Tetraodon, Mugil, Tilapia, Clarias and Malapterurus.   Tomb of Two Brothers, Saqqara, Dynasty V.

2.28 (top) *The seine fishnet. Tomb of Rahotep, Medum, Dynasty IV.*
2.29 (bottom) *Ancient Egyptian fishnets. British Museum, London, and el-Hibeh, Egypt (inset).*

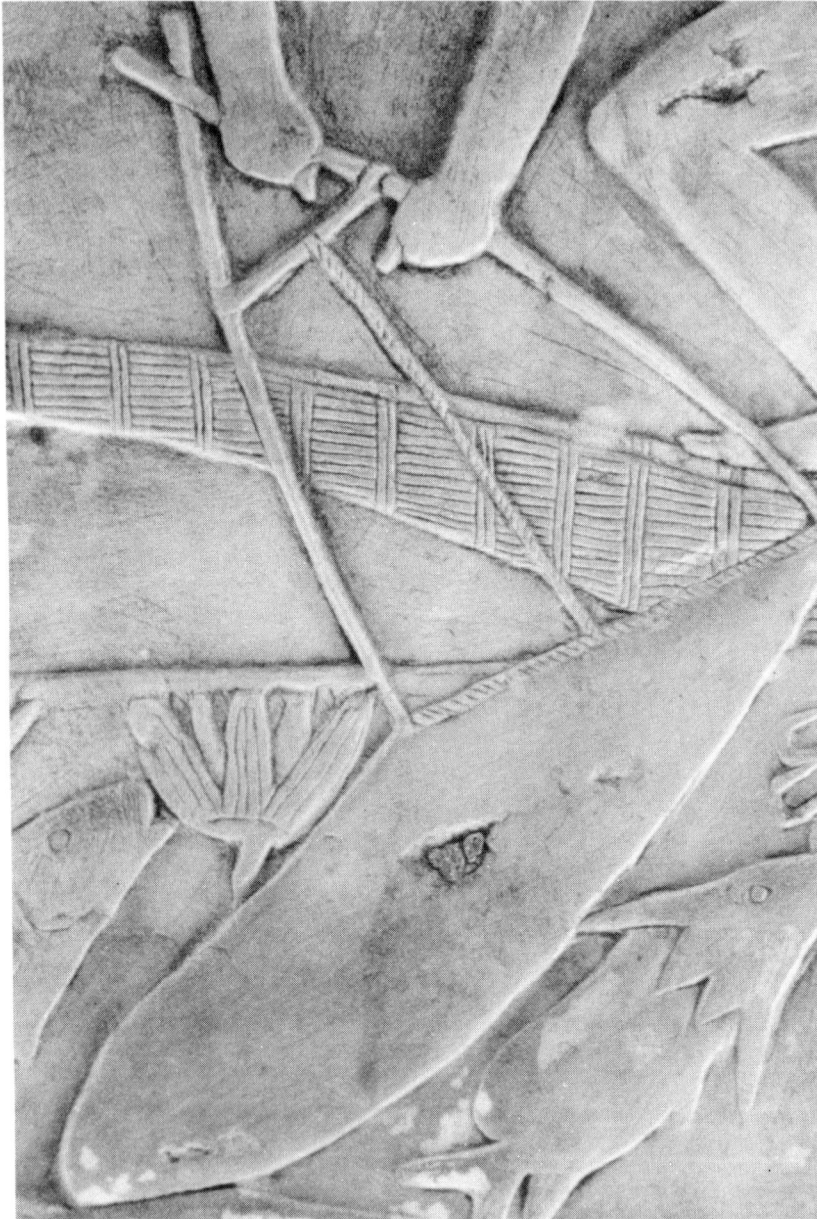

2.30 The common hand net depicted in Egyptian relief. Tomb of Kagemni, Saqqara, Dynasty VI.

longitudinal cord fastened to the terminal cord at the end of the "V".

In some representations the brace is absent and in others the longitudinal cord or drawstring is missing (fig. 2.31). Whether this is an oversight on the part of the artist or whether it depicts variations of the hand net cannot be ascertained. The nets themselves were generally deep and attached to the side sticks. In use, the nets were held by the crotch of the "V" or by the transverse brace, enabling the fisherman to lever up the net, or the net was grasped at the crotch of the "V" and at a point far enough out on one of the side pieces to give similar control (fig. 2.30, 2.31). Hand nets are often depicted in association with

angling and are about as common.[78] In the majority of Old Kingdom examples the hand netter is shown working from a boat often accompanied by an angler. On rare occasions a hand netter is shown standing on the shore or in the water (e.g, Tomb of Two Brothers; Kaemnefret)[79]. Today, hand nets can occasionally be seen in use,[80] but professional fishermen use the large seine or gill net almost exclusively.

Cast Nets

Loat[81] states that the cast net was common throughout Egypt at the turn of the century. The simplest, and according to Loat, the most widely used cast net had an average circumference of

40

2.31 Fishing with the hand net. Tomb of Idout, Saqqara, Dynasty V.

about 15m. with a 1.5cm. mesh. Circumference and mesh, of course, varied a great deal. Attached to the middle of the net was a strong retrieving cord; a thinner cord ran around the circumference. The net, as its name implies, was cast onto the water, spreading out into a circular shape. After landing on the water the net, sometimes aided by small weights or sinkers attached to the circumference cord, sank to the bottom. The net was then carefully retrieved. The popularity of the cast net has waned in recent times, and today it is difficult to find it in use.

Although once a popular means of catching fish in recent times, cast nets are not represented in Egyptian art. A single exception might be the

net depicted in the tomb of Sebeknakht, el-Kab (fig. 2.32). The fisherman is, however, working the net from a boat, presumably to allow his master the opportunity to spear the entrapped fish. It is possible, given this interpretation, that the net is a seine, which has been gathered and partially retrieved, thus consolidating the catch.[82] Some authors believe that cast nets are not depicted due to the difficult nature of showing a fisherman in the act of casting his net.[83] Interestingly, all other forms of fish nets identified from Egyptian scenes are displayed in action. It seems more likely that the cast net simply was not a popular means of catching fish in ancient times.

41

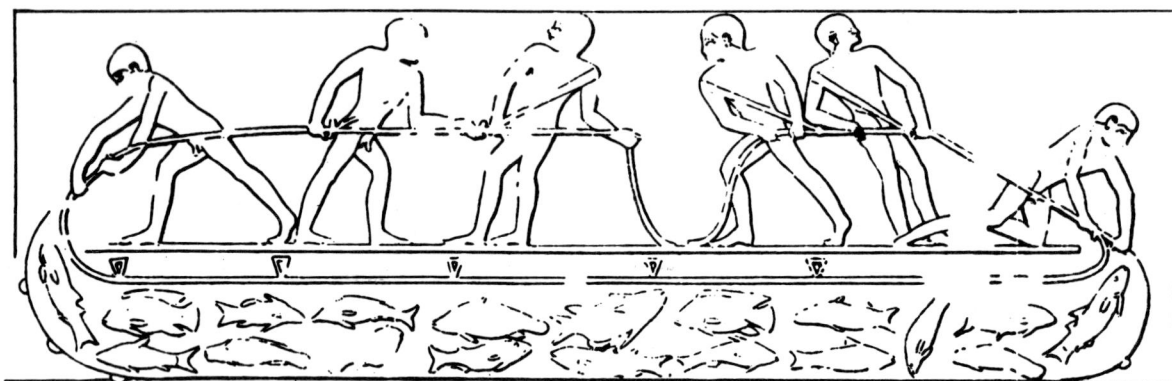

2.32 (top) The cast net (?) of Sebeknakht. Tomb of Sebeknakht, el-Kab, Second Intermediate Period.
2.33 (bottom) Men hauling in a large seine. Tomb of Urarna, Sheikh Said, Dynasty V.

Seine Nets

Fishing with large seine nets is a common theme depicted by Old Kingdom artists.[84] In fact, the earliest artistic representation of a fishing net (fig. 2.28) is that of a seine. The ancient Egyptian seine, as most commonly represented, was a net of considerable length, requiring many men to work it. In shape, it was much like those employed in modern Egypt. It consisted of a long strip of netting with parallel top and bottom support lines and rounded (tapered) ends (fig. 2.33). To each end of the net was attached a harness rope for hauling in the seine (fig. 2.34). To haul the net, the fishermen not only used their hands but frequently employed shoulder slings as well (fig. 2.35). These slings were not made of twisted fibers like the round ropes of the nets, but were flat bands (fig. 2.36). The slings were apparently attached to the harness ropes by wrapping a round ball or knot around the rope. The ball and rope would then, presumably, catch along the bight of the harness rope and jam when a strain was put on it (fig. 2.37).

2.34 (top) *The tapered or rounded end of a fishing seine·showing the attached hauling rope. Tomb of Ti, Saqqara, Dynasty V.*

2.35 (bottom) *A seine being hauled in by men employing the shoulder sling. Tomb of Two Brothers, Saqqara, Dynasty V.*

The bottom line of the seine was weighted to make it hang vertically when in use (fig. 2.38).[85] The weights were originally made of stone (fig. 2.39) and possibly ceramic; in later times lead was also employed.

The upper line of the seine was dressed with floats (fig. 2.38). Representations of floats are, however, so stylized that it is impossible to define their true shape. Nevertheless, where they are more carefully illustrated, the floats appear to be triangular pieces of wood lashed at the end to the net or the rope. The most common type of Old Kingdom float appears in figure 2.38. Floats of this general type lasted, at least in the scenes, into New Kingdom times. Present-day floats are made from plastic jugs, cork, or virtually anything that floats.

Large nets were most often worked from the river or channel bank but, large rafts or boats as shown in figure 2.40 were also employed. To understand this scene it is necessary to refer to the ethnographic record. Egyptian fishermen at the turn of the century, like their modern counterparts, used a boat to work their way slowly across the water while playing out their net. The net is laid in the form of a great arc. Often one of the fishermen will beat the water with his oar to scare the fish into the arc. The fishermen eventually closed the arc after having traversed a

2.36 (top) A close up of a shoulder sling. Tomb of Mereruka, Saqqara, Dynasty VI.
2.37 (bottom) The sling shown with the ball unattached to the harness rope. Tomb of Ti, Saqqara, Dynasty V.

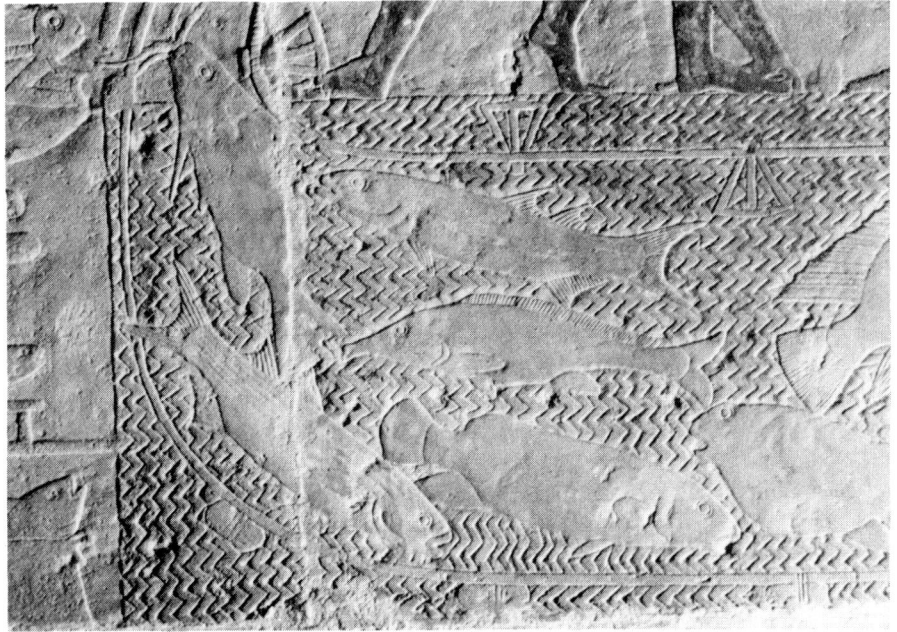

2.38 (top) Representation of Old Kingdom net weights and floats. Tomb of Ti, Saqqara, Dynasty V.

2.39 (centre) A stone net weight. Egyptian Museum, Cairo.

2.40 (bottom) Hauling in the seine. Tomb of Zau, Deir el Gebrâwi, Dynasty VI.

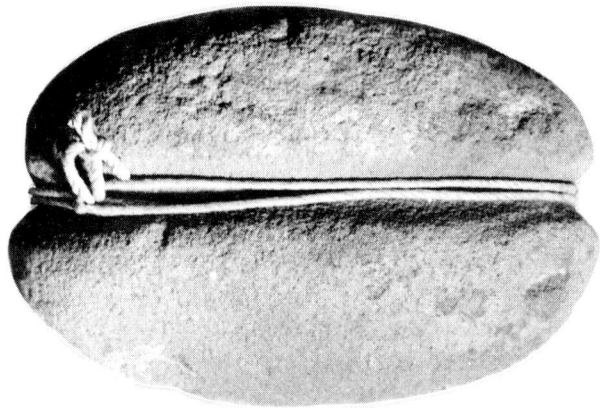

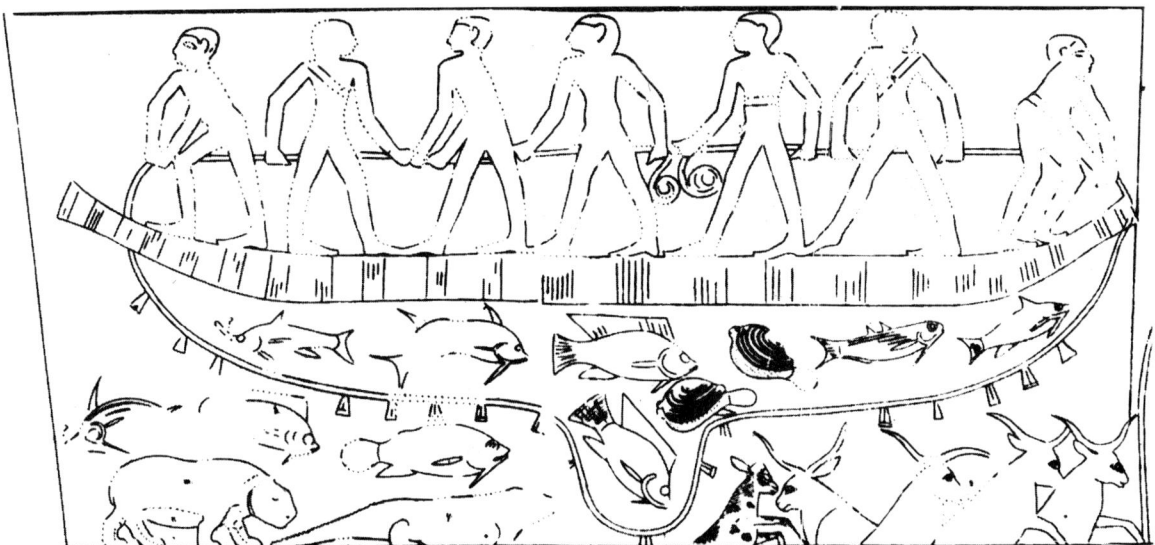

45

full circle.[86] When the bow man can pick up the other end of the seine, the hauling begins: this is the moment the artist has chosen to represent in figure 2.40. Only the top line of the net is visible, and it is obviously out of proportion to the papyrus raft of the fishermen.

Other techniques were also used, as shown in figure 2.41. Here, the fishermen hauling one end of the seine are on the shore; the other end is being manipulated by men in a boat. In figure 2.42 the net is being pulled through the water or otherwise manipulated by two boats.

Netting scenes become progressively more lively throughout the Old Kingdom as more detail and up to 28 people are added to the scenes. Examples include removing the spines from *Synodontis*, adding an overseer, and showing different physical characteristics of the the haulers (those at the ends often being depicted as elderly men).[87] The imagery of the net, which must have been a common sight in ancient Egypt, as it is today, made its way into the religion as exemplified by the Middle Kingdom Coffin Texts. In them, a series of spells are concerned with protecting the deceased from the nets of the underworld demons.[88]

2.41 (top) *Hauling in a seine using men on both land and boat. Tomb of Ahanekht, el-Bersheh, First Intermediate Period.*
2.42 (bottom) *Two boats pulling a seine. Tomb of Meketre, Dynasty XI, Egyptian Museum, Cairo.*

46

FISHES REPRESENTED IN EGYPTIAN ART

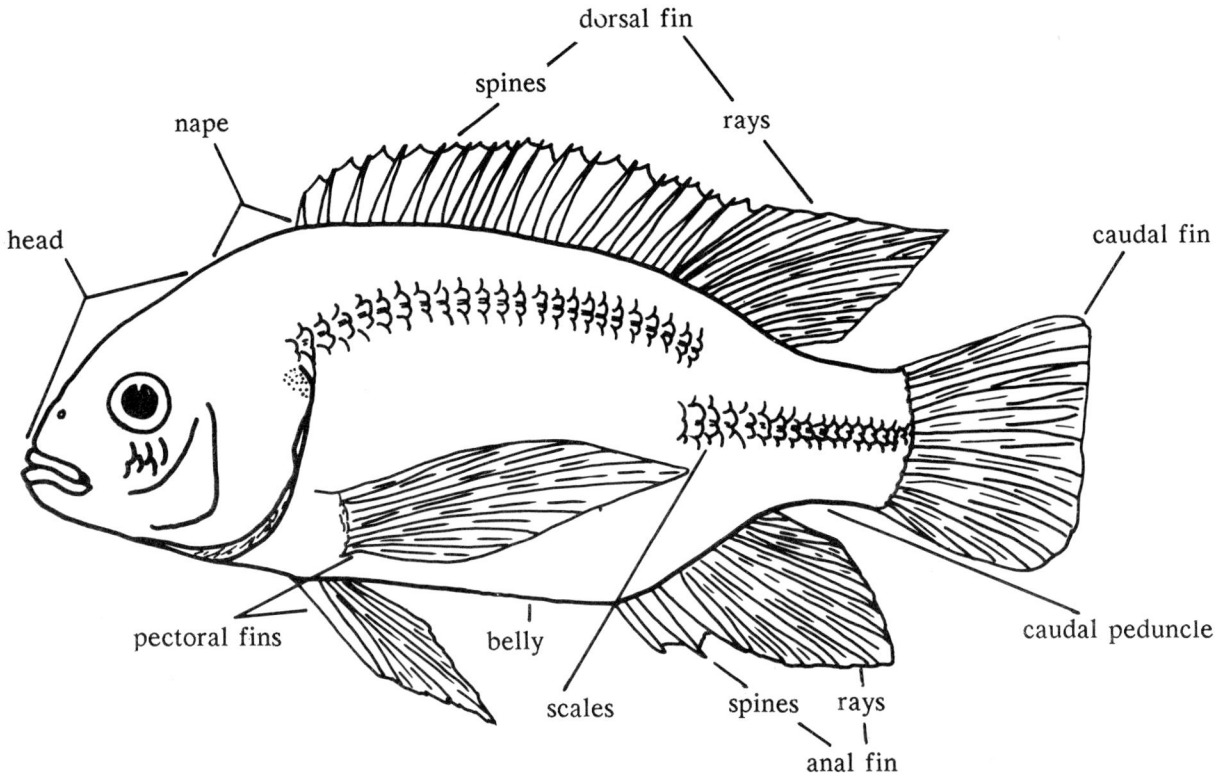

3.1 The important physical characteristics used to identify the fishes of ancient Egypt. Not pictured are barbels (sensory filaments "whiskers" located near the mouth), ventral fins (paired fins located near the belly region) and the adipose fin (a soft fleshy growth located behind the dorsal fin).

Presented in this chapter are the zoological criteria needed to identify the Nile fish portrayed in Egyptian art and hieroglyphs.[1] Representative examples of the piscine taxa are shown to assist scholars in the identification of Nile fish encountered in their own studies. The terminology used to describe the various morphological characters of these fishes, and the artistic representations, are illustrated in figure 3.1. The information provided on the natural history of these fish should stimulate new interpretations of ancient Egyptian fishing scenes and of Nile fish in general.

In many cases, the fish represented in Egyptian art cannot be identified below the level of genus. The zoological criteria used to identify many living fish are dependent upon coloration, the number of scales within a defined area, gill-raker counts, and the number of boney and soft rays making up the dorsal and anal fins. These characteristics are not often shown in the ancient representations, either because of poor preservation (as is most often the case with color), or because the portrayal of the fish did not accurately represent these subtle details.

It is not known to what level the ancient Egyptians could actually recognize differences between fish. In many cases, the criteria used to separate species within a genus involve characteristics that can not easily be discerned in field situations. Even modern Egyptian fishermen, whose livelihood is dependent upon these fish, often confuse some taxa with others because of the subtle nature of many of the defining characteristics. Given the nature of these criteria, it is also likely that the ancient Egyptians could not distinguish between many of the same taxa. Therefore, making a specific identification on the basis of morphological differences in artistic forms not founded in the zoological literature as being taxonomically significant might add an unjustified degree of precision to the fishes portrayed by the ancient artists.

Therefore, the identifications reported herein are based solely on morphological criteria that are expressed on the reliefs or effigies in question. Observed differences between similar forms that cannot be related to significant taxonomic characteristics are not considered as a basis for assigning species epithets.

SUPERCLASS: PISCES

ORDER: ISOSPONDYLI

FAMILY: MORMYRIDAE (elephant-snout fishes)

The family Mormyridae is restricted to Africa. As a group, mormyrids are considered to be bottom feeders and, in general, lack spinous fins.[2] Several members of the family possess long trunk-like snouts from which the common English name for the family "elephant-snout fishes" has been derived. However, snout length and shape vary considerably across and within genera. Snouts range from being short and rounded, as in the genus *Petrocephalus*, to elongate and trunk-like, such as those of the genus *Mormyrus*.

Another interesting characteristic of the family Mormyridae concerns the modified muscles of the caudal peduncle, which form an electric organ. This organ creates a weak electromagnetic field and is believed to function as part of a special electro-sensory system, facilitating the orientation and communication abilities in this group of fish.[3] Studies carried out on the electrical properties of *Mormyrus kannume* show that a continuous stream of impulses is emitted from the organ. When the fish is at rest, discharge frequency is low but rapidly increases to between 80 and 100 impulses per minute when the fish is disturbed. Any electrical conductor entering the electro-magnetic field around the fish evokes an immediate response. The function of the organ is thus analogous to a radar warning device, which is of considerable value to a fish whose life is spent in muddy waters where visibility is poor.[4] The means by which disturbances in the electric field are detected by the fish is still undetermined. Certain neuro-glandular epidermal cells, the "mormyromasts", are possibly associated with the perception of changes in the fish's environment.[5] The level of electrical activity, incidentally, is not enough to give a person a strong shock.

In addition to the electric organ, members of this family possess several other anatomical, osteological, and histological peculiarities. The brain is especially noteworthy for its great size. The brain weight of these fish is approximately 1/52 to 1/82 of the fish's total weight. This is unparalleled among lower vertebrates.[6] Because the greatest degree of hypertrophy is found in the cerebellum and acoustico-lateralis areas of the hind brain, it is not unreasonable to suppose that the functional significance of this area of the brain may be related to the coordination of sensory impulses detected through organs other than the eyes, and therefore indirectly to the presence of the electric organ.

Genus: *Petrocephalus/Marcusenius*
Synonyms: (*Marcusenius*: *Heteromormyrus*)
Egyptian names: *arminya*; *glumaya*; *anooma*

The general identifying characters of the genus *Petrocephalus* include an inferior mouth situated below the eyes, a short body, and ventral fins located nearer to the pectoral fins than the anal fin. Additionally, the dorsal and anal fins are nearly equal in length, the latter being somewhat longer. *Petrocephalus* spp. eat algae, plankton, small fish, and crustaceans. Mature eggs have been recovered from females in August.[7]

Four species belonging to this genus are known to occur in the Nile: *P. bane*, *P. bovei*, *P. keatingii*, and *P. degeni*.[8] All members are relatively small (ca. 12 cm).[9] *Petrocephalus keatingii* is known from only a few specimens collected in the Sudan[10]; *P. degeni*, from a single specimen caught in Lake Victoria.[11] *Petrocephalus bovei* is reported to be the most abundant mormyrid in the Lower Nile.[12] Few specimens of this genus, however, were encountered when information was being gathered for this volume. Additionally, Latif[13] reports that very few specimens were caught in Lake Nasser during the Lake Nasser Fisheries Survey.

Petrocephalus bovei is often confounded with *P. bane* and *Marcusenius isidori* by local Egyptian fishermen. The three closely related taxa can only be reliably separated by the number of dorsal fin rays and by tooth characteristics. Unfortunately, these characteristics were not

48

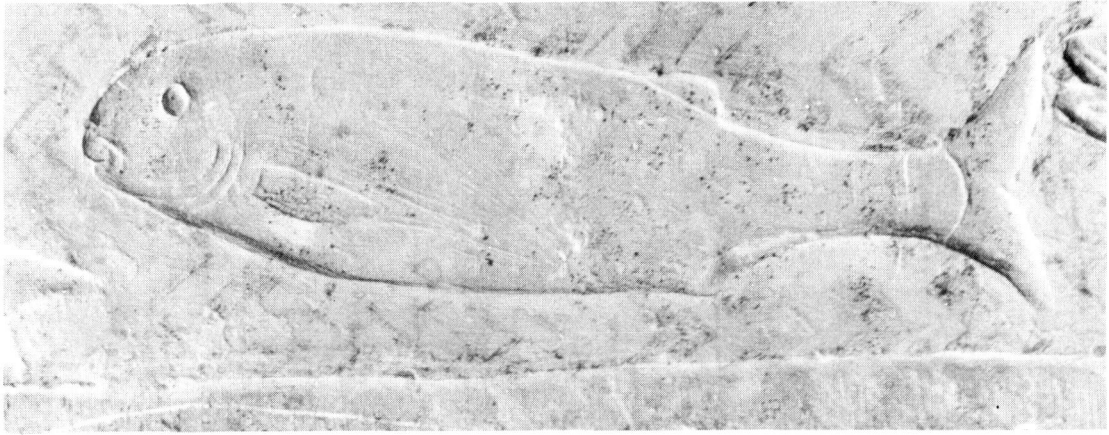

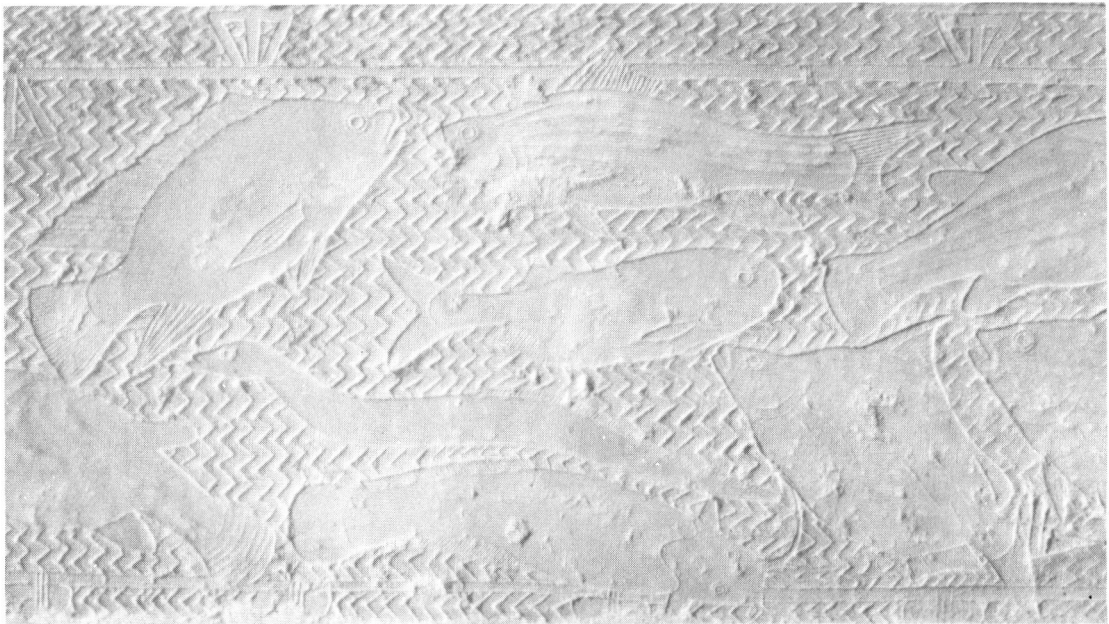

3.2 (top) An early Dynasty VI example of Petrocephalus/Marcusenius. Tomb of Mereruka, Saqqara.
3.3 (bottom) An example of Petrocephalus/Marcusenius from the Tomb of Ti, Saqqara, Dynasty V.

portrayed in the ancient representations (figs. 3.2, 3.3). Consequently, the identifications of mormyrids possessing characteristics resembling that of *Petrocephalus* are here referred to as *Petrocephalus/Marcusenius*. Such a conservative identification is warranted given the nature of the characteristics used to distinguish the actual living taxa.

49

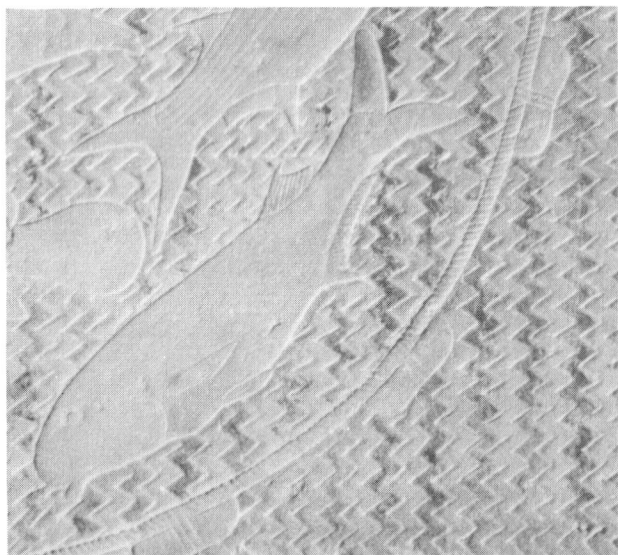

Genus: *Gnathonemus*

Synonyms: *Campylomormyrus*

Egyptian names: *anooma; om-shifefa*

Five members of this genus have been recorded from the Nile, but only one species, *Gnathonemus cyprinoides*, has a distribution that includes the Egyptian Nile. Thus, this taxon would appear to be the only referent for the ancient Egyptian representations of this fish.

Gnathonemus cyprinoides has been recovered from the Delta to the Bahr el-Gebel, Sudan. It is, however, more numerous in the Delta areas of Egypt than in Upper Egyptian waters, such as Lake Nasser, where few specimens have been caught.[14] *Gnathonemus cyprinoides* feeds mainly on underwater insects and spends a great deal of time in aquatic vegetation.[15] The preference for aquatic vegetation may explain the wide distribution and preference for the Delta. This fish is thought to be nocturnal.[16]

Gnathonemus cyprinoides resembles *Petrocephalus* in many ways but has a more elongate body and a mouth in the terminal position.[17] Additionally, the ventral fins of this fish are equidistant from the anal and pectorals, and the dorsal and anal fins are nearly equal in length. The caudal peduncle is, in contrast to *Petrocephalus*, much thinner and the caudal fin terminates in pointed lobes. Probably the single most distinguishing characteristic of *G. cyprinoides*, however, is its swollen chin.[18]

In comparing the representations of *Petrocephalus*-like fishes, a subtle distinction can be made in the body length of two of the representations (figs. 3.4, 3.5). Additionally, these two representations show a slightly different form for the mouth. In both cases the mouth is more ventrally located and the lower jaw is slightly extended or enlarged. Although none of the fish represented show an exaggerated or swollen chin, the two representations possess consistent traits. Consequently, an identification of cf. *G. cyprinoides* can be suggested. To identify this taxon securely, it would be necessary to identify similar covarying traits in other scenes from other tombs or localities. Only by such comparisons can we be sure that the traits represented are taxonomic characteristics and not products of artistic expression or variation.

3.4 (top) The swollen chin of cf. Gnathonemus cyprinoides. Tomb of Ti, Saqqara, Dynasty V.
3.5 (bottom) An elongated form of cf. Gnathonemus cyprinoides. Tomb of Ti, Saqqara, Dynasty V.

Genus: *Mormyrus*

Synonyms: *Scrophicephalus*; *Mormyrodes*; *Solenomormyrus*.

Egyptian name: *boueza*.

This genus is represented in the Nile by four species: *Mormyrus kannume*, *M. caschive*, *M. niloticus*, and *M. hasselquistii*, although only two species (*M. kannume* and *M. caschive*) are recovered with any frequency. *Mormyrus kannume*, *M. caschive*, and *M. niloticus* cannot be mistaken for any other fish. These species possess an elongate snout; the Egyptian vernacular name, *boueza*, refers to this characteristic. *Mormyrus hasselquistii*, on the other hand, lacks the long characteristic snout of this genus. In 1907, Boulenger stated that *M. hasselquistii* had not been recovered since Geoffry St. Hilaire collected the original four specimens near Cairo.[19] Although specimens of this taxon are now available for study, it is still considered rare.

Members of this genus, in general, possess an elongate body with a long dorsal fin. The eye is usually tiny. *Mormyrus kannume* and *M. caschive* can be separated from *M. niloticus* by the curvature of their snouts. In *M. niloticus* the snout is elongate and straight; in the other two species the snout curves downward.

Although *M. kannume* generally dies soon after being taken into captivity, Flower[20] noticed that aquarium specimens were very active at night, spending the day quietly on the bottom of the aquarium. Also, *M. kannume* was observed swimming backwards; leading with the tail. *Mormyrus kannume* and *M. caschive* feed predominantly on insect larvae, particularly larval lake flies (Chironomidae)[21], but are also known to eat fish, earthworms, and bottom animals.[22] These fish can be caught in offshore and near-shore waters but are less frequently encountered in areas of dense vegetation.[23] Gonads are mature in spring and summer, but only a left ovary or testes is present. Sexual maturity, apparently, is not attained until these fish reach a size of about 30 cm.[24] Specimens up to 1 meter in length have been reported for *M. caschive* and *M. kannume*.[25]

The local fishermen do not generally recognize *M. kannume* as being distinct from *M. caschive* because the differences between these fish are quite subtle. They are best separated on the basis of their dorsal fin rays: *M. caschive* has significantly more rays (76-90) than *M. kannume* (57-75).[26] A good field identification can still be made, however, on the basis of the type of snout and the location of the dorsal fin. In *M. caschive* the snout is more slender than in *M. kannume*, and the dorsal fin in *M. caschive* originates midway between the pectoral and ventral fins. In *M. kannume* the dorsal fin originates almost even with the ventrals.[27] In *M. niloticus* the dorsal fin

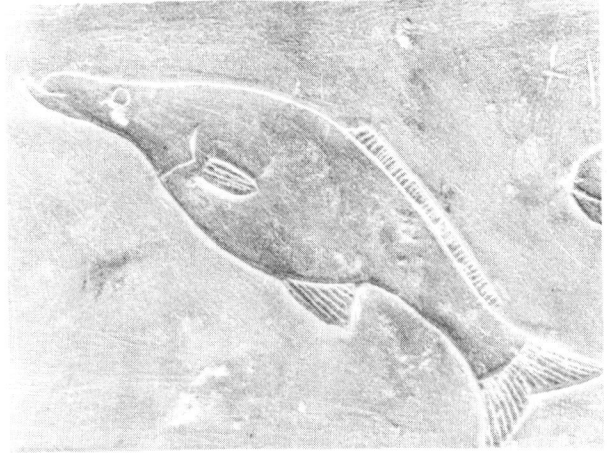

3.6 (top) *Relief of Mormyrus cf. kannume emphasizing its thick snout. Tomb of Idout, Saqqara, Dynasty V.*

3.7 (bottom) *Mormyrus cf. cashive, which possesses a slender curved snout. Tomb of Mereruka, Saqqara, Dynasty VI.*

51

appears in the same location as in *M. caschive* but the snout is straight rather than curved.

Given the fact that modern fishermen cannot distinguish between *M. kannume* and *M. caschive*, it seems possible that the ancient Egyptians could have confused these taxa as well. There is, however, evidence to the contrary. The ancient Egyptians appear to have been able to distinguish between members of the genus *Mormyrus* and attempted to portray this fact by altering the size and shape of the fish's snout. For this reason,

representations that possess an obviously thick snout have been assigned to the taxon *M. cf. kannume* (fig. 3.6). Those representations showing a thin but strongly curved snout are assigned to *M. cf. caschive* (fig. 3.7). Figures possessing a thin straight snout are referred to as *M. cf. niloticus* (fig. 3.8). The dorsal fin, although an important zoological characteristic, appears to vary considerably across representations, rendering it of little use for purposes of identification.

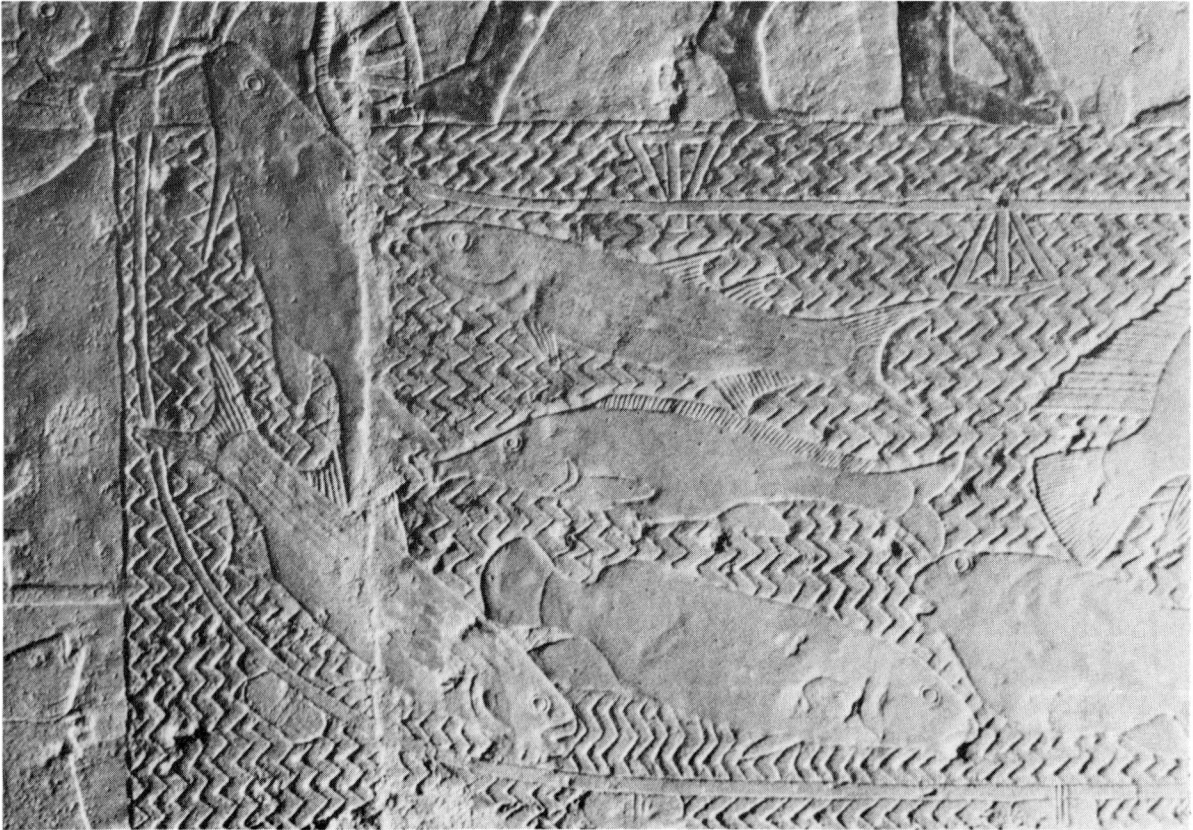

3.8 A likeness of Mormyrus cf. niloticus, identified by its strait snout. Tomb of Ti, Saqqara, Dynasty V.

Genus: *Hyperopisus*

Synonyms: *Phagrus*

Egyptian names: *sawiya, galmier*

One species belonging to the genus *Hyperopisus* is widely distributed in the Nile (*H. bebe*). It is reported to prefer fluvial conditions[28], although Daget[29] mentions that *H. b. occidentalis* is common in swamps and other areas inundated during seasonal flooding of the Niger. This fish is also known to feed on molluscs and water insects.[30] Additionally, it is said to be fond of the millet that spills off the Nile grain barges.[31]

There are numerous field characteristics that can be used to identify this taxon: the head is scaleless, somewhat rounded, and is slightly longer than deep; the mouth is small, terminal, and situated below the horizontal level of the eye. This genus can most readily be distinguished, however, by its small dorsal fin and long anal fin.[32] The fish has a strongly curved upper profile with a dorsal fin placed far from the head, near the caudal fin. The anal fin originates mid-body at a point equidistant from head and tail.[33]

Only two possible referents for this fish can be identified from the ancient representations (figs. 3.9, 3.10). The two examples come from the Tomb of Mereruka. Although it appears, on the basis of comparisons to other representations of mormyrids, that the ancients were indeed attempting to represent a different type of fish, it is difficult to assess the diagnostic importance of the features presented. Without comparisons to figures from other localities, any single characteristic thought to be diagnostic could just as reasonably reflect local artistic expression. Nevertheless, the position of the dorsal fin so close to the tail, the ventrally located gill slits, general body form, and the location of the mouth suggest that these representations might be *Hyperopisus bebe*. This identification should be considered tenuous, however. Therefore, a conservative identification of cf. *H. bebe* is recommended.

3.9 *(top)* *From the Tomb of Mereruka, an example of cf. Hyperopisus bebe. Saqqara, Dynasty VI.*

3.10 *(bottom)* *A boating scene representing cf. Hyperopisus bebe. Tomb of Mereruka, Dynasty VI.*

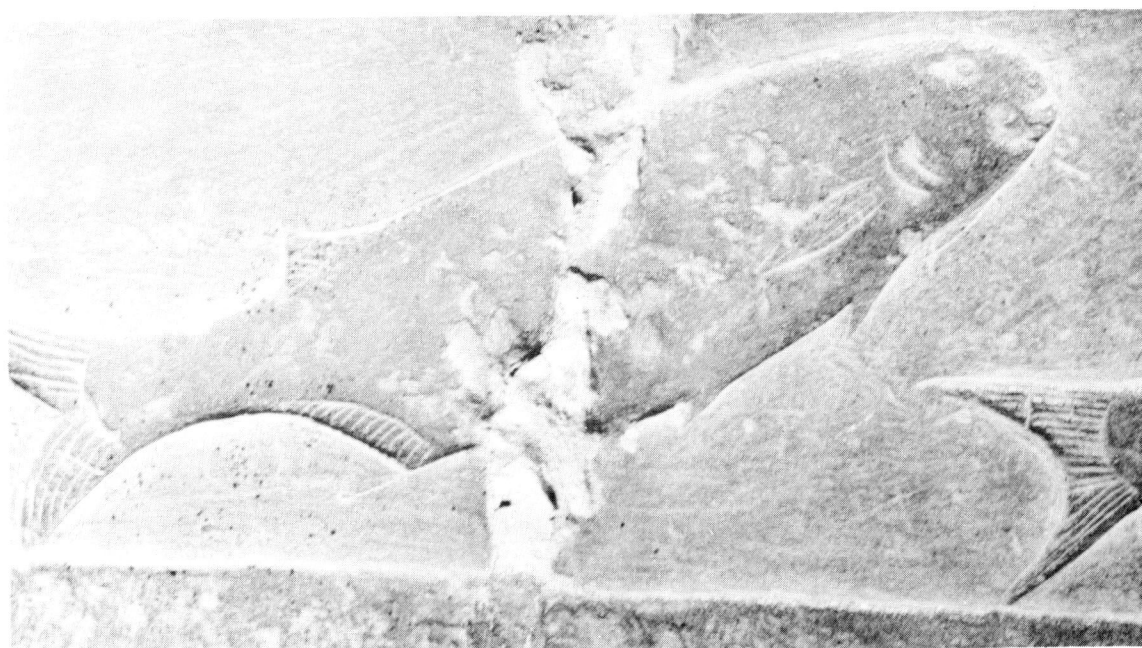

ORDER: OSTARIOPHYSI
FAMILY: CHARACIDAE (tiger-fishes)

Members of this family are characterized by strong teeth, especially *Hydrocynus* spp. All East African species possess an adipose fin.

Genus: *Hydrocynus*

Synonyms: *Hydrocyon, Hydrocinus*

Egyptian name: *kelb el-bahr*

The genus *Hydrocynus* is characterized by a scaleless head, sharply edged teeth, elongate snout and body, rounded belly, a small adipose fin, and nostrils that are close together and near the eyes.[34]

Three species have been recorded from the Egyptian Nile; two are common, and the third is considered rare. *Hydrocynus forskalii* has been recovered from all parts of the Nile system; *H. brevis*, from Luxor and the Lower Nile; *H. lineatus*, only from Lake Nasser. *Hydrocynus lineatus* is rarely encountered in Egypt because its northern distribution extends only to the southern portion of Upper Egypt. Lake Nasser lies along the very northern periphery of its range.[35]

The three species of *Hydrocynus* found in the Nile can be separated from one another on the basis of body shape, subtle differences in the location of the dorsal fin, and scale counts. *Hydrocynus forskalii*, which is the most common species in the Egyptian Nile, possesses a more slender body and is lighter in color than the other two taxa. In addition to being thicker bodied, *H. brevis* and *H. lineatus* are also more boldly marked than *H. forskalii*.[36]

Adult specimens of this genus studied by Boulenger[37] ranged in length from 26 to 60 cm. *Hydrocynus* can be found near inshore waters and in offshore waters down to a considerable depth.[38] It is most often recovered during the annual flood.[39] *Hydrocynus* spp. feed mostly on fish, although insect larvae and crustaceans have been noted in stomach contents of this genus.[40] In Lake Nasser during the winter, *H. forskalii* is known to feed mainly on insect larvae. During the other months of the year it feeds on fish, insects, and shrimp. Stomach contents have included fish of the genus *Alestes, Mormyrus, Labeo, Tilapia, Synodontis,* and *Eutropius,* as well as smaller members of its own genus. On the basis of a year-long study of *Hydrocynus* stomach contents, *Alestes* is the most frequent food fish, followed by *Tilapia (Sarotherodon).*[41]

Hydrocynus is the most likely candidate for the Phagrus of ancient Egyptian legends.[42] It is interesting that at only one ancient site is it represented. Representations of *Hydrocynus* were identified by Keimer[43] from the tomb of Ankhtifi at Mo'alla near Esna (fig. 3.11).

3.11 A rare representation of Hydrocynus. Tomb of Ankhtifi, Mo' alla, First Intermediate Period. See Pl. VII.

54

Pl VII fig 3.11 (top) Hydrocynus (centre) with Tilapia (l. below) and a Synodontis (lower l.). Tomb of Ankhtifi, Mo'alla, First Intermediate Period.

Pl VIII fig 3.25 Tilapia (top), Synodontis (centre) and Mugil (bottom). Tomb of Two Brothers, Saqqara, Dynasty V.

Pl IX fig 3.28 Mugil, Tilapia, Clarias, Malapterurus & Citharinus (top to bottom, l. to r.). Tomb of Two Brothers, Saqqara, Dynasty V.

Pl X   fig 3.26 (top) Synodontis. Tomb of Seankhuptah, Saqqara, Dynasty VI.
Pl XI   fig 3.30 (centre) The Egyptian eel. Dashur, Dynasty VI, Egyptian Museum, Cairo.
Pl XII   fig 3.39 (bottom) A pond scene with Tilapia. Nebamun's Garden, Thebes, Dynasty XVIII, British Museum, London.

Genus: *Alestes*

Synonyms: *Myletes, Brycinus, Brachyalestes*

Egyptian names: *sardina, kelb el-bahr, raya*

Members of the genus *Alestes* are similar to *Hydrocynus* in general form and appearance, but are smaller and feed predominantly on insects rather than other fish.[44] *Alestes* are characterized by short snouts; nostrils near the eyes, an elongate and compressed body, rounded belly, and moderate to large scales. The dorsal fin of this genus is behind the ventral fin and, like all East African characids, *Alestes* possesses a small adipose fin.[45] The teeth of this genus are multicuspidate and fixed in two rows.[46]

Boulenger[47] has reported five species of *Alestes* in the Nile system. Three species have been recovered from the Egyptian Nile: *A. dentex, A. baremose*, and *A. nurse*. These taxa can be separated on the basis of fin ray and gill raker counts as well as a subtle difference in the location of the dorsal fin. This genus usually does not grow to an excessively large size. Although *A. baremose* can reach 40cm. in length, *A. nurse* and *A. dentex* generally do not exceed 12cm. in length. *Alestes* with mature ovaries have been observed throughout most of the year in Lake Nasser.[48] *Alestes* stomach contents have included insects, small fish, bottom debris, snails, and plant fragments.[49] *Alestes* frequently travel in small schools near the surface of the water and often jump, especially in the evening.

The only clear representations of this fish are found at the temple complex of Medinet Habu (fig. 3.12). The rounded belly and small adipose fin set it apart from other taxa.

3.12 An example of Alestes sp. from a scene depicting Ramses III. Medinet Habu, Luxor, Dynasty XX.

FAMILY: CITHARINIDAE (moon-fishes)

Fish of this family are easy to identify by their distinctive shape. The body is comparatively deep and these fish possess a conspicuous adipose fin.[50] Four genera have been recorded from the Nile, but only two, *Distichodus* and *Citharinus*, have been recovered with any frequency. The less common genera, *Nannaethiops* and *Nannocharax*, are small fish that live near inshore vegetation along the Upper Nile.[51] They can be separated from other tiny fish found in the same habitat by the presence of an adipose fin. *Citharinus* is the only member of this family that can be identified among the ancient artistic representations.

Genus: *Citharinus*
Egyptian names: *amara, gamer*

The genus *Citharinus* can be identified by its short snout, wide terminal to sub-inferior mouth, strongly compressed body, and a conspicuous adipose fin.[52] The deep body and concave nape serve as the best identifying field characteristics. Two species of this genus have been identified in Egypt: *C. citharus* and *C. latus*. *Citharinus latus* can be separated from *C. citharus* on the basis of the length of the adipose fin. In *C. citharus* the adipose fin (measured at the base) is shorter than the distance from adipose fin to the rayed dorsal fin. The base of the adipose fin in *C. latus* is longer than the distance between the rayed dorsal and the beginning of the adipose fin.[53] *Citharinus citharus* possesses a falcate dorsal fin; *C. latus* possesses a rounded dorsal fin.

Citharinus spp. was at one time common throughout the Nile; it has been recorded in the Lower Nile, Upper Nile, Blue Nile, and White Nile.[54] It has, unfortunately, become relatively rare in recent times.[55] Members of this genus feed on macroplankton, particularly Crustacea and diatoms that settle on the river bottom.[56] *Citharinus citharus* has been seen swimming on the surface, apparently feeding on algae.[57] No information on *Citharinus* breeding habits is available for the Egyptian Nile. In the rivers Niger and Gambia, spawning occurs in swampy areas during the heavy autumn rains.[58]

On the basis of the length of the adipose fin, artistic representations more closely resemble *Citharinus citharus*, but a considerable amount of variation exists in the artistic portrayal of this feature (fig. 3.13, 3.14). Dorsal fins from these representations, either falcate or rounded, do not consistently correlate with adipose fin length. Variations in the the concave area of the nape also fail to accurately represent the two respective species.

In summary, although it appears likely that the ancient Egyptians were aware of the varying morphological characteristics within this genus, it is difficult to assign a species epithet to the representations because of the inconsistency in the depiction of diagnostic features. Consequently, it would seem prudent to refer to the representations as *Citharinus* spp. According to Loat[59], modern Nile fishermen also only recognize *Citharinus* at the generic level. Therefore, to assign a species epithet to the artistic forms might only serve to add an unfounded degree of precision.

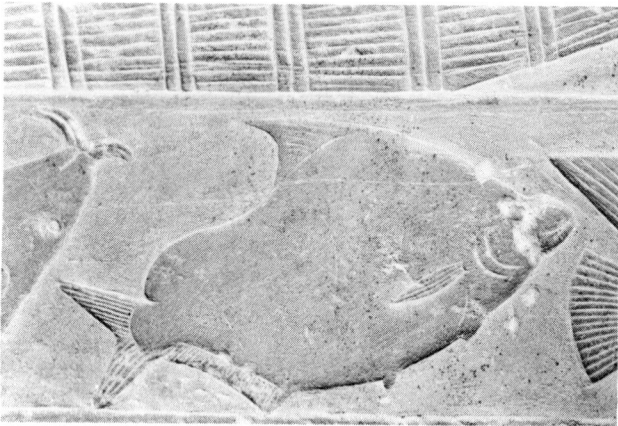

3.13 (top) A representation of Citharinus sp. Tomb of Mereruka, Saqqara, Dynasty VI.
3.14 (bottom) A second form of Citharinus sp. Tomb of Mereruka, Saqqara, Dynasty VI.

FAMILY: CYPRINIDAE (carps)

The family Cyprinidae is widespread; its members inhabit most of the worlds freshwaters. The most characteristic feature of the family is its pharyngeal teeth. This characteristic, unfortunately, offers little assistance in the identification of the artistic representations. Besides the pharyngeal teeth, cyprinids can be distinguished by the position of the ventral fins, a forked caudal fin, large scales, and the lack of an adipose fin.[60] Five genera have been recorded in the Nile. Two genera, *Labeo* and *Barbus*, contain members of sufficient size to be of economic importance. All other members of this family are small fish having little economic value.[61] *Labeo* and *Barbus* are the only two genera that have been identified in the ancient artistic representations.

Genus: *Labeo*

Synonyms: *Abrostomus, Tylognathus, Rohitichthys.*
Egyptian names: *lebis, debs.*

The genus *Labeo* is widespread and has been recovered throughout Africa and southern Asia.[62] As a whole, the genus is best characterized by its well developed lips, which form a ventral sucker. The lips vary in shape according to species and are thus important identifying characteristics.[63] Barbels are usually present near the corners of the mouth[64], and the mouth itself is somewhat crescent shaped. The snouts of some members of this genus have well defined horny tubercles, especially among males during the breeding season. The scales of this genus possess longitudinal (head to tail) stria. Additionally, several species possess indistinct longitudinal lines (head to tail) on the lower flanks of their body. Five species of *Labeo* have been identified from the Nile, and four

3.15 The only recognizable example of the genus Labeo. Tomb of Ti, Saqqara, Dynasty V.

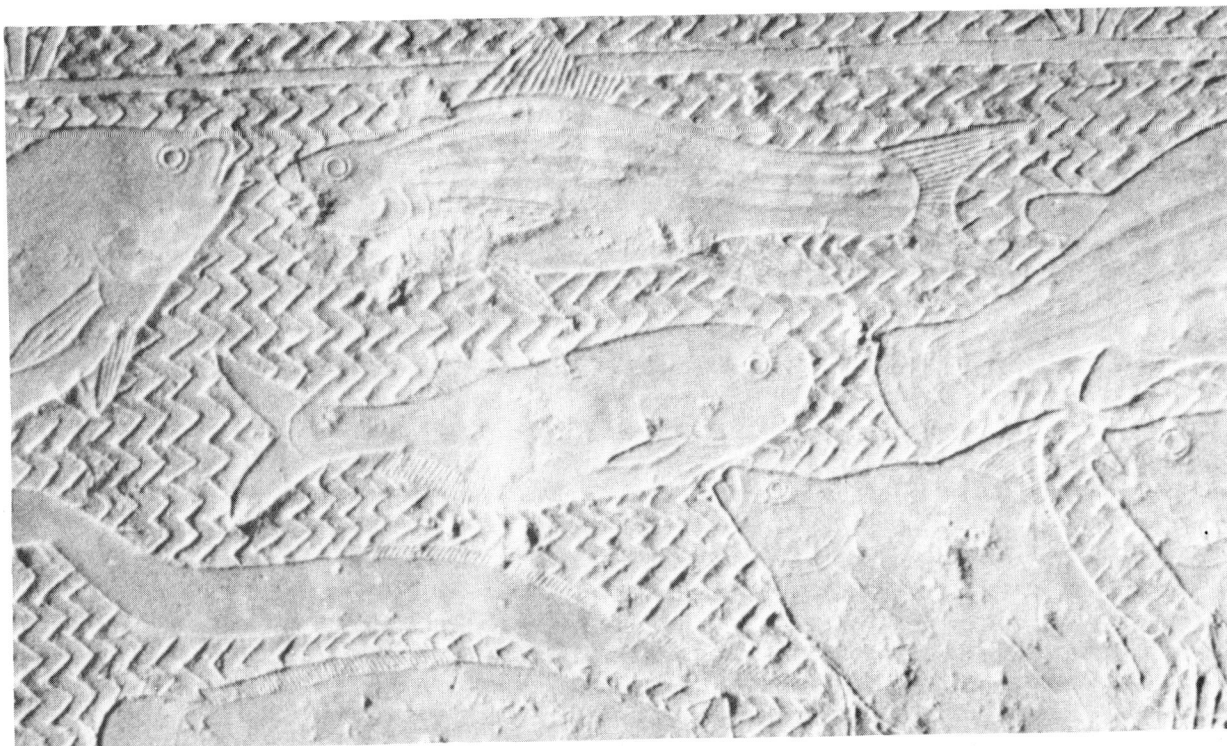

57

species are unique to the Nile. *Labeo coubie, L. niloticus, L. horie,* and *L. forskalii* have been recorded from the Egyptian Nile. All four taxa principally inhabit inshore waters and reach sizes in excess of 50cm. in length.[65]

Labeo niloticus is by far the most common *Labeo* in the Egyptian Nile.[66] The genus *Labeo* contributed 24% of all the fish produced from the Nile in 1965 and 18% of the fish recovered in 1966. The predominant species was *L. niloticus.*[67] It is the most common *Labeo* found in the northern areas of Egypt, but it is less frequently encountered in southern Egyptian waters. *Labeo horie* and *L. coubie,* on the other hand, are rare in the Lower Nile but are relatively more abundant in the southern waters of Upper Egypt. *Labeo forskalii* is said to occur over the entire Nile system but is mainly confined to rocky areas.[68]

Fish of this genus are essentially herbivorous bottom feeders. They are known to swallow great quantities of mud from which they digest the diatoms and other minute organisms living in the sediments.[69] Stomach contents have also included leaves, grasses, algae and, to a lessor extent, detritus and insects.[70]

According to Loat[71], *L. horie* and *L. niloticus* are not recognized as separate taxa by the Nile fishermen who were also not able to distinguish any difference between *L. forskalii* and *L. coubie.* Latif[72] reports that *L. coubie* and *L. horie* were not distinguished as separate types of fish by the local fishermen of Lake Nasser. Confounding these taxa with each other is not surprising, for ichthyologists separate these four species on the basis of the number of branched dorsal fin rays, presence or absence of tubercles, and the presence or absence of a rostral flap.

Only one representation of the genus *Labeo* can be positively identified from the ancient works.[73] It comes from the Tomb of Ti at Saqqara (fig. 3.15). The profile of this fish looks much like the representations of *Barbus* cf. *bynni,* with one important exception. The flank of this representation displays distinct longitudinal stripes. The ancient artists were obviously distinguishing this cyprinid from the others on the basis of these stripes, which refer to either the striated scales or the longitudinal stripes found on such common taxa as *Labeo niloticus.* A closer inspection also shows that the dorsal fin of this figure is not as strongly curved as those depicted on *Barbus* and that its posterior border is free of the body. A specific identification, however, is impossible for several reasons. Most importantly, no other complete representations of this taxon have been reported. This limits the identification to the generic level. Specific identifications require several representations to compare and contrast the many varying characteristics. Such comparisons are essential in determining which characteristics are taxonomically significant and which can most likely be attributed to artistic expression.

Genus: *Barbus*

Synonyms: *Labeobarbus*, *Cheilobarbus*, *Pseudobarbus*, *Capoëta*, *Systomus*, *Puntius*, *Enteromius*, *Barynotus*.

Species: *B. bynni*

Egyptian name: *bynni*

The genus *Barbus* includes approximately 1500 species spread throughout Asia, Africa, and Europe.[74] More species are assigned to *Barbus* than any other freshwater non-cichlid genus in Africa.[75] Intraspecific variability in this genus, however, is extremely high and many so-called species are probably geographic variants of the same species.[76] Greenwood[77] has presented a revision of some East African species. Unfortunately, the list does not include many Nilotic forms equally in need of revision.

Barbus spp. are generally thought to be omnivorous.[78] Sandon and Tayib[79], however, believe these fish to be mainly herbivorous. Identifying characteristics for *Barbus* spp. include a terminal mouth and large scales. A predominant characteristic of the more common species is an ossified dorsal fin spine. Smaller members of this genus generally lack an ossified dorsal spine and possess smaller scales.

Representations of *Barbus* are common in Egyptian art and have been identified in tomb reliefs and effigies. Assigning species epithets to these likenesses does, however, require an involved process of elimination. Boulenger[80] lists 35 species of *Barbus* in the Nile system. Fortunately, most of these fish are very small[81] and can easily be discounted as referents to the artistic forms. The list can be further reduced by the fact that many ancient representations display a strong dorsal spine and the spine is not serrated. Only one species displaying this trait has been frequently recorded from the Egyptian Nile, *Barbus bynni*. Although *B. werneri*, *B. neglectus*, *B. perince*, and *B. anema* have been recovered from the Nile and Lake Nasser[82], all lack a strong dorsal spine. All four taxa are small fish that possess distinct lateral markings, which aid in separating them taxonomically from young *B. bynni*.[83] *Barbus bynni* is thus the most likely candidate for the ancient artistic representations of a *Barbus*-type fish and they are so classified (fig. 3.16, 3.17).

Barbus bynni has a wide distribution. Its range extends from the Delta to the White Nile. It has been taken from Lake Manzala during the flood season and is fairly common in the Bahr el Yousef. It is even known to occur in irrigation ponds and trenches. *Barbus bynni* is best distinguished by its single dorsal spine, high humped nape, and a deeply forked caudal fin. Additionally, *B. bynni* possesses two barbels on each side of its mouth. Mature ovaries in fish

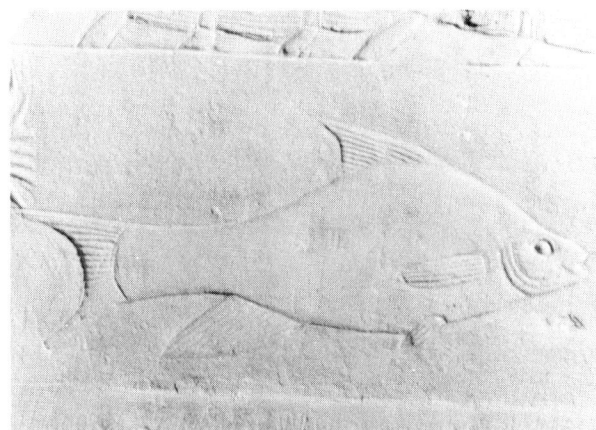

3.16 (top) *Barbus bynni* (higher fish) from the tomb of Idout, Saqqara, Dynasty V.
3.17 (bottom) *Barbus bynni* from the tomb of Mereruka, Saqqara, Dynasty VI.

taken from Lake Nasser have been observed during the spring and summer months.[84] In the Nozha-Hydrodrome, on the Nile Delta near the Mediterranean Sea, *B. bynni* spawns from April to September[85]; some are able to reproduce as early as late February.[86] *Barbus bynni* prefers clear inshore waters and thrives in either riverine or lacustrine environments.[87] This fish can attain a relatively large size, rendering it economically important. Loat[88] recorded specimens of *B. bynni* in the Fayum reaching nearly 0.5m. long and in Cairo a specimen 0.65m. long, weighing 5.5kg. has been recorded.

59

SUBORDER: SILUROIDEA (catfishes)
FAMILY: CLARIIDAE

The family Clariidae is restricted to the Nile system, Africa, and Southeast Asia. It has recently been introduced to North America, where it is known as the walking catfish. Family characters include a much depressed head, four pairs of barbels (one nasal, one maxillary, and two mandibular), small eyes, and a very elongated dorsal and anal fin.

This family is represented in Egypt by two genera, *Clarias* and *Heterobranchus*; each is represented by two species. *Clarias* and *Heterobranchus* can be easily separated from one another by the presence or absence of an adipose fin. In *Heterobranchus* the dorsal fin is divided in two parts: a rayed dorsal fin followed by an adipose fin. The genus *Clarias* has only one rayed dorsal fin that extends over nearly the entire length of the fish.

Genus: *Clarias*
Synonyms: *Macropteronotus*,
Egyptian names: armoot, garmoot, hoot.

Two species of *Clarias*, *C. lazera*, and *C. anguillaris*, are known to inhabit the Egyptian Nile.[89] It should be noted, however, that the *Clarias* species of Africa are in urgent need of taxonomic revision.[90] Intraspecific variability is high and many so-called species will undoubtedly prove to be merely geographic variants of a single widespread species. Some revisions have been made[91], although none based on Nilotic species have been reported.

The hardiness of *Clarias* is legendary among the fishermen and villagers of Egypt. Although the stories are undoubtedly exaggerated, there is no doubt that *Clarias* can tolerate conditions that would kill most other fishes. *Clarias* prefer shallow deoxygenated waters.[92] They have also been recovered from relatively deep waters with a muddy bottom. The ability of *Clarias* to survive in foul stagnant water is due to an elaborate accessory breathing organ that enables these fish to use atmospheric oxygen. The accessory organs are in the form of highly branched bodies developed from two of the gill arches on each side of the fish. These bodies are contained within an expansion of the gill chamber, which is lined with a highly vascular skin and protected by the bones of the skull. The tissue covering the organ and the skin lining the suprabrancial chamber have the same microscopic structure as the normal gill filaments.[93] It is across these surfaces that oxygen is absorbed from the air. Even though its gill filaments are well developed, it appears that *Clarias* is dependent on the additional supply of oxygen for survival. Fish prevented from gaining access to the surface drown.[94]

The amphibious habits of *Clarias* may explain its ability to rapidly populate new areas. There are many records from various parts of Africa of *Clarias* moving overland from one body of water to another; it has been seen leaving the water and moving overland during the day and night.[95] Welman[96] personally observed a column of about 30 *C. lazera* move overland from a shallow swamp to the back-water of a river. The total distance travelled was about 200m. and it took nearly one hour for the fish to cover this distance. When temporary streams and swamps dry up, *Clarias* are reputed to burrow in the mud and remain there until the area is again flooded. Such behavior, however, has yet to be substantiated and warrants further study.

Clarias possesses a single long dorsal fin as well as a single elongate anal fin. Dorsal and anal fins are composed entirely of soft rays. The outer margin of both pectoral fins, however, is armed with a sharp ossified spine. The upper and lateral parts of the head of this fish are osseous and covered with coarse to fine bumps. These small bumps can be easily seen through the thin covering of skin.

The two species of Nile *Clarias* are best separated on the basis of gill-raker counts. *Clarias anguillaris* has 20 to 27 gill rakers; *C. lazera*, the most common Nile *Clarias*, possesses 35 (young) to 135 gill rakers[97]. Both species readily attain sizes in excess of 1m. in length. A good, but not always reliable, field characteristic that can be used to separate these two taxa is the presence of a pale, colored edge on the caudal fin of *C. anguillaris*. *Clarias lazera*, in contrast, possesses a uniformly colored caudal fin.[98] Additionally, *C. anguillaris* possesses a slightly coarser granulated pattern on its skull than does *C. lazera*. However, this characteristic is variable. The post-orbital bone has long been considered a taxonomic trait useful in separating these two taxa, but this characteristic is also quite variable.[99]

Nile fishermen believe that *Clarias* spp. comes out of the water onto the ground to eat vegetation[100]; however, no records substantiate this. Studies conducted by Pekkola[101] show that *Clarias* feeds on molluscs and fish. One specimen examined by Pekkola contained algae in its stomach, but no specimens contained any terrestrial plant material. Sandon and Tayib[102] report that specimens taken near Khartoum contained only fish remains. One large specimen examined by Sandon and Tayib contained mollusc shell fragments, fish scales, crustacean appendages, and some mud containing diatoms. *Clarias* collected near Cairo by Brewer contained only fish remains sometimes mixed with mud.[103]

Although *Clarias* spp. are generally considered to be predacious fish, *C. lazera* is also able to filter minute animals and plants from the water by means of its long, closely set gill rakers.[104] The compiled evidence on the diet of *Clarias* seems, consequently, to indicate that they would be best described as omnivorous. Stomach contents have included insect larvae, molluscs, planktonic organisms, water weeds, and fish such as *Tilapia* (*Sarotherodon*).[105] Poll[106] recovered a *Tilapia* 35cm. long from the stomach of a large *C. lazera*.

Fish of this genus can be collected by hand, particularly during the spawning season.[107] There are, however, no published records of when spawning occurs for *C. lazera* or *C. anguillaris* in the Nile. The spawning season is thought to correspond with the annual period of inundation occurring in the late summer, but specimens of *C. lazera* collected near Luxor have been found to contain mature eggs as early as April.[108] Greenwood[109], however, notes that *C. mossambicus* (now *C. gariepinus*), a species whose relationship to *C. lazera* is in need of investigation (some authors consider *C. lazera* a variant of *C. gariepinus*), breeds in small temporary streams that flow into Lake Victoria during the spring rainy season in April and again during the winter rains in December.

Evidence appears to indicate that *Clarias* may be an opportunistic breeder. Spawning has been triggered in some *Clarias* species by artificially raising the water level of the lakes in which they live. When conditions become favorable, spawning can commence within 36 hours.[110] Two or as many as three episodes of spawning can occur during a single season, depending on environmental conditions. It is

likely that no single physical or chemical factor is responsible for triggering the spawning of *Clarias* species. Changes in water level, duration of daylight hours, chemistry, temperature, water clarity, flow velocity, and biological factors such as the flooding of marginal plants and access to suitable spawning sites may all be important factors conditioning spawning activities.[111]

The ancient representations of *Clarias* are unique in that the head of this fish is always shown as being viewed from above, but the body is figured in profile (fig 3.18). *Clarias* and *Heterobranchus* are the only fish represented in this fashion. It is interesting to note that when working with an actual specimen, it is difficult to position the head to display a profile view. The shape of the head and body tend to offer only a dorsal view unless some means is used to anchor the entire fish in profile.

On the basis of the representations, *Clarias* can only be identified at the generic level. Modern Egyptian fishermen do not recognize any difference between the two zoological species, and variations in the ancient artistic representations do not correspond to any significant taxonomic characteristics. Some authors have attempted to assign species epithets to the artistic forms on the basis of the shape of the post-orbital bones, but this characteristic is variable in the artistic scenes and in the actual zoological specimens (fig. 3.19, 3.20). Other characteristics that vary across figures are the number of barbels and the placement of the dorsal fin in relation to the anal fin. Because these features are not zoologically diagnostic, it is impossible to know if these observed differences are the result of the artists' attempts to portray different taxa or are simply insignificant variations.

3.18 An example of Clarias sp. Dashur, Dynasty VI, Egyptian Museum Cairo.

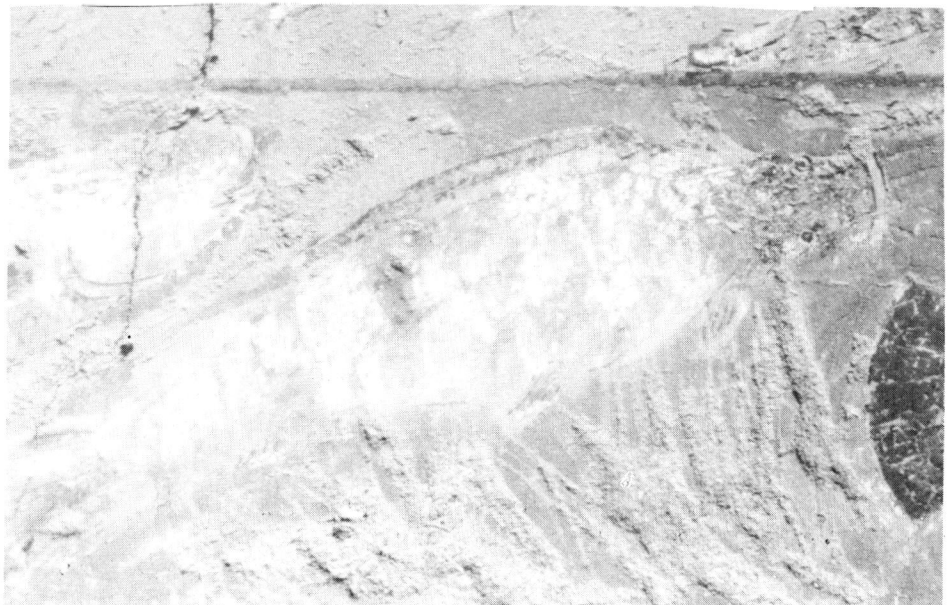

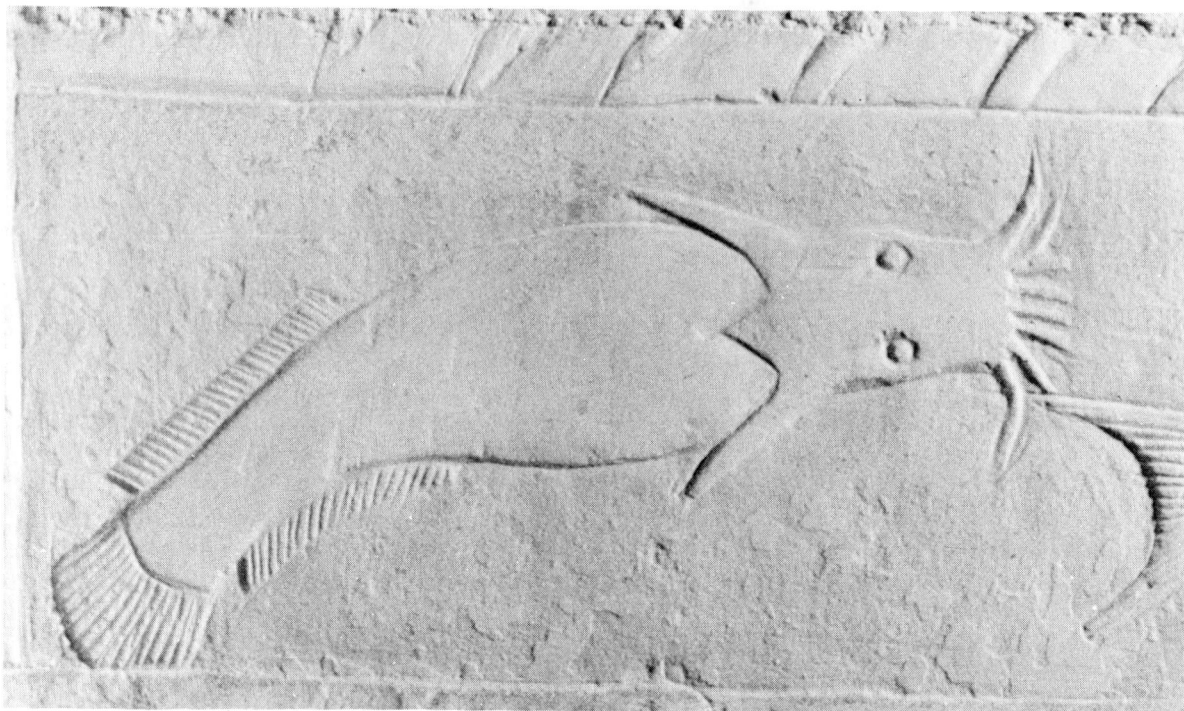

3.19 (top) A representation of Clarias sp. depicting a sharply pointed post-orbital oone. Tomb of Mereruka, Saqqara. Dynasty VI.

3.20 (bottom) Clarias sp. with a rounded post orbital bone. Tomb of Mereruka, Saqqara, Dynasty VI.

Genus: *Heterobranchus*
Egyptian names: *armoot, garmoot, hoot*

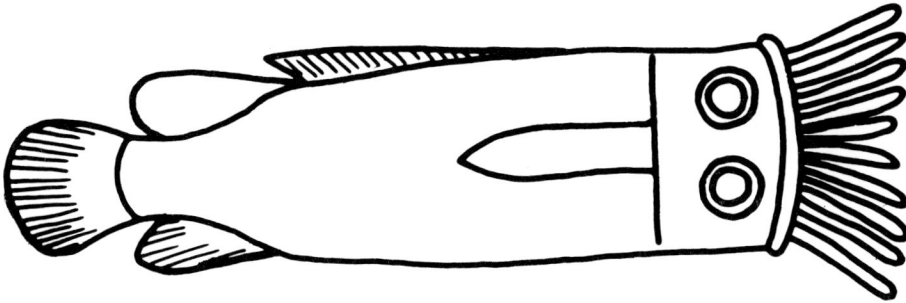

3.21 *Heterobranchus as depicted on the Narmer palette.*

Heterobranchus is often confounded with *Clarias* by the local fishermen. Two species of *Heterobranchus* occur in the Nile; they can be separated by the relative length of their adipose fin. *Heterobranchus bidorsalis* has a short adipose fin and a long dorsal fin composed of 38 to 45 soft rays. *Heterobranchus longifilis* has a long adipose fin and a shorter dorsal fin composed of 29 to 34 soft rays. *Heterobranchus bidorsalis* is found predominantly in southern Upper Egypt and the Sudan; *H. longifilis* has been recorded from the Nile near Luxor. Latif has recorded both species in Lake Nasser.[112]

Unfortunately, not much is known about this genus. *Heterobranchus* is believed to be a bottom feeder and thought to possess the amphibious-like qualities of *Clarias*.[113] Although *Clarias* has been recovered from deep waters by trummel and sinking gill nets, Worthington[114] also recovered *Heterobranchus* from the turbulent water below the Murchison Falls. It may be that *Heterobranchus* prefers waters with some current vis-à-vis *Clarias*, which thrives in the more stagnant waters of lacustrine and riverine environments.

Among the earliest identifiable representations of a fish in Egypt is that of *Heterobranchus*. The image of what appears to be a *Heterobranchus* can be found on the famous Narmer palette recovered from Hierakonpolis (fig. 3.21).[115] The dual dorsal fin and catfish-like whiskers limit the possible referents for this representation to the genus *Heterobranchus* or possibly *Bagrus*. The rounded caudal fin and lack of a prominent dorsal fin spine further limit the possibilities to include only the genus *Heterobranchus*. The exact identity of King Narmer is in question. Nevertheless, most scholars agree that the palette is of late Predynastic date. Consequently, with the exception of the pisciform palettes recovered from Predynastic graves, *Heterobranchus* is one of the oldest identifiable artistic representations of a fish in Egypt.

FAMILY: SCHILBEIDAE

Members of the family Schilbeidae are well known to the fishermen of the Nile. They can be easily distinguished from the Clariids by their short dorsal fin and barbels. Additionally, their caudal peduncle appears to a varying degree bent downward, giving the fish its characteristic humped or bent appearance. This family is represented in Egypt by four genera: *Physailia*, *Eutropius*, *Schilbe*, and *Siluranodon*. These taxa can be easily separated by the presence or absence of an adipose fin, dorsal spine, and humped nape. *Schilbe* is the only genus belonging to this family represented in Egyptian art.

Genus: *Schilbe*
Egyptian name: *schibla*

The genus *Schilbe* can be easily recognized in the ancient representations by its slightly projecting lower jaw, humped nape, and the presence of dorsal and pectoral fin spines. Further diagnostic characteristics include the lack of an adipose fin (a characteristic that separates this genus from *Eutropius*) and a deeply forked caudal fin.[116] *Schilbe* is featured prominently in many Old Kingdom tomb scenes and in carved or cast effigies. Additionally, *Schilbe* is the only fish used by the ancient Egyptians as a nome standard. It is the referent for the "Mendes fish".

Two species of *Schilbe* are potential candidates for the artistic representations: *S. mystus* and *S. uranoscopus*. *Schilbe mystus* is a common fish of the lower Nile. It eats anthropods, diatoms, and fish, particularly *Alestes*.[117] Mature gonads of this fish have been observed in the summer. *Schilbe uranoscopus* is thought to possess a more southerly distribution than *S. mystus* because it is relatively more numerous in the Blue Nile and White Nile. It should be cautioned, however, that the type specimen for *S. uranoscopus* comes from Cairo, although it is not very abundant in the Cairo area today. *Schilbe uranoscopus* like other members of the family, is a surface fish and feeds mostly on insects. Other foods include crustaceans and small

3.22 *A representation of Schilbe sp. emphasizing the pectoral and dorsal spines. Tomb of Mereruka, Saqqara, Dynasty VI.*

64

fish. The gonads of *S. uranoscopus* are ripe in the summer; in August spent females are frequently encountered.[118]

The easiest way to separate these two taxa is by reference to the humped nape of *S. uranoscopus*. Unfortunately, this characteristic is not one that is easily distinguished on the ancient representations (fig. 3.22, 3.23). Given that an unknown degree of artistic license was taken by the ancient artists, features such as a greater hump in the nape on one representation compared to another might not be taxonomically important. Variations in the nape of other fish such as *Lates niloticus*, which can be identified on the basis of other characteristics, show that body proportions are not always constant from scene to scene and such characteristics should not be used singly as the basis of an identification. In such cases a suite of characteristics are needed. In the case of *Schilbe* we have conflicting evidence concerning the characteristics needed to make a specific identification. *Schilbe mystus* would seem the most likely candidate because it is more common in the Egyptian Nile. Yet the representations always show a very humped nape, a characteristic most diagnostic of *S. uranoscopus*. Consequently, without additional significant biological traits to use as a basis for identification, it seems best to remain conservative and refer to representations of this taxon as simply *Schilbe* sp.[119]

3.23 *The Schilbe standard of Mendes. Late Period, Egyptian Museum Cairo.*

65

FAMILY: BAGRIDAE

The family Bagridae is well represented in the Egyptian Nile. Although four genera have been recovered from the Nile (*Bagrus, Chrysichthys, Clarotes,* and *Auchenoglanis*), only one (*Bagrus*) has been identified from the ancient representations. The family name is derived from this genus and it is a well known taxon to the Egyptian fishermen.

Genus: *Bagrus*
Synonyms: *Silurus, Porcus.*
Egyptian names: *bayad, docmac*

Bagrus can be easily separated from other catfish by its long adipose fin, deeply forked caudal fin, dorsal and pectoral fin spines, and a head that is quite depressed. Two species of *Bagrus* are known to inhabit the Egyptian Nile: *B. docmac* and *B. bayad.* They can be easily separated on the basis of two field characteristics. The easiest characteristic for identification is the presence on *B. bayad* of a long, slender extension on each lobe of the caudal fin; *B. docmac* has an extension on only the upper lobe of its caudal fin.[120] The second field characteristic is the position of the dorsal fin. In *B. docmac* the last ray of the dorsal fin is in front of the last ray of the pelvic fin; in *B. bayad* the last ray of the dorsal fin is behind the last ray of the pelvic fin.[121] A third characteristic, based on the bone structure of the skull, can be used to separate these taxa, but it cannot be used to identify artistic representations of the fish because artists rendered this taxon in profile. Loat[122] states that the Nile fishermen had separate names for *B. bayad* (bayad) and *B. docmac* (docmac or farfour). Evidence compiled for this study indicates, however, that modern fishermen cannot distinguish the two taxa as separate species or, at best, recognize the existence of different kinds of *Bagrus* but refer to them all by the same name.

Bagrus is a predatory fish that feeds mainly on insect larvae, shrimp, and small fish[123]; molluscs have also been found in *Bagrus* stomach contents.[124] Egyptian fishermen working the Nile for *Bagrus* generally fish in the deeper waters where some current can be detected.[125] In Lake Nasser, however, *B. bayad* can be caught in shallow water with rocky bottoms. From this information and from personal observations it seems that the genus may prefer water possessing some current rather than simply shallow or deep water. Aquarium studies have shown that *Bagrus* hides during the day and is active at night.[126] *Bagrus* is, therefore, thought to be a night feeder. *Bagrus docmac* and *B. bayad* can attain sizes in excess of 1m. long.[127]

In Lake Nasser, spawning nests were seen in shallow waters with a sandy-loamy bottom near mountain areas or large rocks. Egyptian fishermen say that one of the parents keeps watch over the nest, swimming out if disturbed, to drive away intruders.[128] Near Beni Suef, *B. bayad* reportedly breeds in January using holes along the banks of the river to deposit its eggs.[129] In the Sudan spawning occurs in May.[130] Recent evidence has shown that although ovaries of *B. docmac* have been found to be mature in summer, eggs of many different sizes were recorded, denoting a fractional or opportunistic spawning characteristic for this fish.[131]

One representation of *Bagrus* has been identified from ancient Egypt from a tomb at Meir (fig. 3.24). It would appear from the projections on both lobes of the caudal fin that the ancient artist(s) were attempting to depict *B. bayad.* This identification could be substantiated if other scenes existed for comparison. In lieu of a larger comparative sample, a taxonomic classification of *Bagrus* sp. is suggested, although an identification of *Bagrus* cf. *bayad* would be acceptable if accompanied by a proper explanation or citation.

3.24 *A rare example of the genus Bagrus. Tomb of Ukh-Hotep, Meir, Dynasty XII.*

FAMILY: MOCHOKIDAE

This family is represented in Egypt by four genera (*Chiloglanis*, *Synodontis*, *Andersonia*, and *Mochocus*), but only the genus *Synodontis* is common[132], and only *Synodontis* is represented in Egyptian art. The most distinctive characteristic of this family is the presence of a cephalo-nuchal shield, which is prominently displayed on the ancient representations.

Genus: *Synodontis*

Synonyms: *Brachysynodontis, Pseudosynodontis, Hemisynodontis, Leiosynodontis.*
Egyptian names: schall, gargoor.

The genus *Synodontis* is confined to tropical Africa and the Nile. This is one of the most important fish of the Nile in number of species and abundance of individuals. A total of thirteen species have been identified from the Nile system.[133] Members of the genus are easy to recognize by the presence of a cephalo-nuchal shield and very strong dorsal and pectoral fin spines. Another related, but significant, characteristic for field identification is the presence of the post-humeral process.

In members of this genus a boney casque (cephalo-nuchal shield) is formed by the union of the skull bones with the nuchal shield. This union is possible due to an expansion of the anterior interneural bones and the clavicle bones. The cephalo-nuchal shield can be easily recognized on live specimens as well as in the ancient representations. Additionally, part of the cephalo-nuchal shield complex, the post-humeral process, can be used for speciating members of this genus.[134] This characteristic is also prominently displayed in *Synodontis* representations.

The sharp, heavily serrated pectoral spines of this fish articulate with the boney casque by means of a complicated joint located on the cleithrum. This joint is arranged so that the spine can be firmly locked into an erect position.[135] The serrated spines offer a formidable defense and can inflict painful wounds. Nile fishermen are aware of the dangers these spines can pose and sometimes use wire cutters to remove the dorsal and pectoral spines before taking these fish from the nets. The ancient Egyptian fishermen also were aware of the danger the spines could pose and took similar precautions in their removal. This fact is attested to in many scenes where the spines of freshly caught *Synodontis* are shown being removed (fig. 2.34).

Although the spines appear to provide a powerful mechanism for defense, young *Synodontis* are often found among the stomach contents of predatory fishes. All members of this genus make sounds produced by the muscles surrounding their air bladder and by creaking the dorsal and pectoral spines.

Nile fishermen have mixed feelings regarding *Synodontis*. The Lake Nasser fishermen are said to dislike the fish because it causes damage to their nets.[136] On the other hand, the fellaheen (peasant) fishermen of the Egyptian Nile are happy to include *Synodontis* among their daily catch; *Synodontis* is regularly eaten by the peasant class.[137]

Habitat preferences of *Synodontis* vary according to species. Most species prefer deep waters, although they are recovered from inshore areas as well. Nile fishermen in Cairo, el-Minya, and Luxor report catching more *Synodontis* in offshore fishing forays vis-à-vis fishing near to shore. Near-shore catches seldom produce *Synodontis*; *Tilapia* (*Sarotherodon*) is the predominant taxon recovered in near-shore catches.[138]

Synodontis spp. possesses an elongate intestinal canal with numerous convolutions. This characteristic, accompanied by a feeble dentition, indicate a diet of vegetable matter and small animals. These fish also show a predilection for decomposing organisms. Loat and Boulenger[139] have found large fish in the stomachs of *Synodontis*. Recent work on *Synodontis* feeding habits demonstrates that differences in diet can be related to the size and age of *Synodontis*.[140] Young fish predominantly feed on insect larvae, algae, and molluscs; mature and thus larger fish feed on a wider variety of organisms. On the basis of this information, *Synodontis* would best be described as an omnivorous bottom feeder.[141]

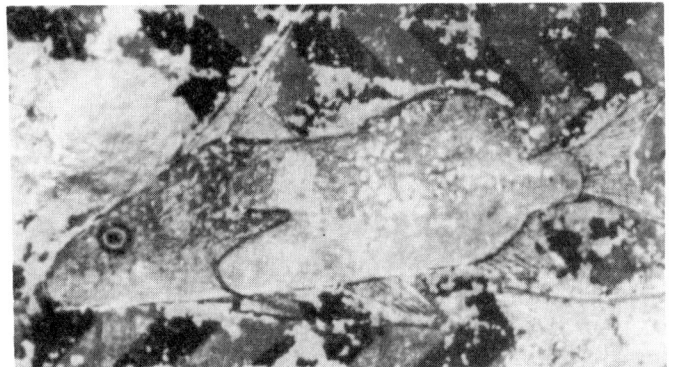

3.25 *Synodontis cf. schall from The Tomb of Two Brothers. Saqqara, Dynasty V. See Pl. VIII.*

67

3.26   A Dynasty VI representation of Synodontis cf. schall from the tomb of Seankhuptah, Saqqara. See Pl. X.

Four species could be possible referents to the artistic representations. Two species, *Synodontis schall*, and *S. batensoda*, can be inferred from the ancient representations. *Synodontis schall* can be easily identified by the pointed shape of the post-humeral process. The post-humeral process of *S. serratus* possesses a more rounded superior border; the anterior border of the dorsal spine is heavily serrated. *Synodontis clarias* is rare in the Lower Nile; only one specimen having been collected. Loat[142] recorded this taxon as living only in the White and Blue Nile. Additionally, this taxon possesses a very short and compressed humeral process with a dorsal spine that is finely serrated on the anterior basal portion and heavily serrated on the posterior border.

Because of the rounded post-humeral process and serrated dorsal spine of *S. serratus* and the rarity of *S. clarias*, the most likely candidate for the artistic representations of *Synodontis* would be *S. schall* (fig. 3.25, 3.26), the most common member of this family. The identification, however, can only be inferred from the

representations. A considerable amount of variability exists in the ancient depiction of the post-humeral process, and evidence seems to indicate that the modern Nile fishermen do not distinguish *S. schall* from *S. serratus*. Therefore, a more conservative identification of *S.* cf. *schall* is recommended.

An interesting fact that has been reported by several naturalists concerns a curious habit of *S. batensoda* and a closely related species, *S. membranaceus*. Both taxa are reported to have a predilection for swimming upside-down. This habit has been noted by Geoffroy and verified by Loat.[143] Studies on the feeding habits of *S. batensoda* and *S. membranaceus* reveal that they are surface feeders. Because their mouths are ventrally located, they must swim upside-down to feed from the surface.[144] When frightened, the fish reverts to the normal position and quickly escapes. Correlated with the unique characteristic of swimming upside-down is a reversal of normal coloration. The true dorsal side has become lightly pigmented and the true ventral side heavily pigmented.[145] Observations of these fish in

68

3.27 Synodontis cf. batensoda from the Tomb of Mereruka, Saqqara, Dynasty VI.

captivity have shown that when food on the water's surface becomes depleted, the fish turn down and feed off the bottom.[146]

During the course of research for this volume, a specimen fitting the description of *S. batensoda* was examined in Luxor. The local fellaheen possessing the fish begrudgingly handed it over for a quick inspection. The fish was heavily pigmented on its ventral surface and lightly colored on its back, indicating that it might be *S. batensoda*. The only other possible referent for the fish would have been *S. membranaceus*. The Egyptian, unfortunately, allowed only a brief inspection of the fish before retrieving it, remarking how it was important for him to return to his home quickly to make a medicine from it.[147]

Because the fish depicted in the ancient scenes appear to be swimming upside-down, it is possible that the ancients knew of this curious habit and attempted to represent it (fig. 3.27). Unfortunately, the fish shown in the reverse position are nearly identical to the *Synodontis* depicted in the normal upright position. It is unfortunate that the color on these scenes has

disappeared because this might have offered clues to their true identities. The presentation of two projections behind the boney casque does not appear to be a significant characteristic because both upright and upside-down representations are shown with this feature. The upside-down representations are, however, more compressed in length than those shown in the normal position. Additionally, a subtle difference can be detected in the adipose fin. *Synodontis* in the normal position have a proportionally smaller adipose fin than those depicted upside-down. Both these characteristics, although quite subtle, are consistent with the actual fish. Furthermore, the ancient Egyptian artists seldom portrayed fish freely swimming upside-down. *Synodontis* and a few instances of the genus *Tilapia* (*Sarotherodon*) are the only known examples. Consequently, there is reason to believe that two taxa are being represented. Upside-down *Synodontis* possessing a slightly larger adipose fin and compressed body are thus referred to as *S.* cf. *batensoda/ membranaceus.*

69

FAMILY: MALAPTERURIDAE (electric catfishes)

3.28 (top) *Malapterurus electricus from the Tomb of Two Brothers, Saqqara, Dynasty V. See Pl. IX.*

3.29 (bottom) *An example of the electric catfish from the Tomb of Ti, Saqqara, Dynasty V.*

Malapteruridae is a monotypic taxon in Egypt, the only member being *Malapterurus electricus*, the electric catfish.

Genus: *Malapterurus*
Species: *M. electricus*
Synonyms: *Raja, Silurus, Malopterurus,*
Egyptian names: *ra'ad, ra'ash.*

Malapterurus electricus, although it ranges over the entire Nile, is not a commonly encountered taxon. It can occasionally be observed, however, in the catches of Nile fishermen. The arabic name, *ra'ad*, literally translates as thunder, but according to the Nile fishermen, it refers to the quivering that results for those who touch it.[148] *Malapterurus* is easy to identify by its cylindrical shape and somewhat bloated appearance (fig. 2.28, 2.29). It possesses a moderately developed adipose fin, rounded caudal fin, and relatively small eye.[149] The skull is somewhat flattened as in most catfishes, and the lower jaw projects slightly. Specimens as large as 70cm. have been caught in Lake Nasser.

This taxon is best known for its electric organ, which is related to the integumentary system, and consists of rhomboidal cells just beneath the skin. These cells extend from the head to the origin of the adipose and anal fins. All the cells are apparently controlled by a single ganglionic cell at the anterior extremity of the spinal cord. In contrast to other electric fish, the current in *M. electricus* proceeds from the head to the tail.[150]

The function of the electric organ is unknown. It undoubtedly serves as a powerful means of defense but also may be used to stun and immobilize the fish's prey.[151] The force of the shock from a mature specimen is said to be around 200 volts.[152] Boulenger[153] cites an interesting account of how a *Malapterurus* used its electric powers to obtain food while in captivity. According to Boulenger, a *Malapterurus* living in the Gizeera aquarium would shock its tank mate, *Clarias lazera*, and eat the earthworms regurgitated by the *Clarias.*

Accounts regarding the preferred habitat of this fish are conflicting. According to Greenwood[154], this species can be found in reed-beds flanking fast flowing water. On the other hand, Latif[155] states that it prefers to live near rocky areas in shallow water with a muddy or sandy/loamy bottom. Egyptian fishermen have provided yet another conflicting account.[156] They say that the *ra'ad* can be caught in deep as well as shallow water, near the shore or away from it, regardless of vegetational cover or current. In dismay they report that the fish can come from anywhere along the Nile; however, it is least often encountered in the winter.

ORDER: ANGUILLIFORMES
FAMILY: ANGUILLIDAE (eels)

Genus: *Anguilla*
Species: *A. vulgaris*
Synonyms: *Muraena*
Egyptian names: *taban samak*

Only one species of eel lives in Egyptian waters, *Anguilla vulgaris*. Although it can be readily observed in the Delta lakes and has been recorded as far south as the second cataract, *A. vulgaris* is relatively rare south of Cairo. The Egyptian eel can be easily identified in the ancient representations by its serpentine body, long dorsal and anal fins, well developed pectoral fins, and well developed jaw (fig. 3.30, 3.31). This is the same eel that is found in the Mediterranean Sea and along the Atlantic coast of Europe.[157]

In Egypt, as elsewhere, the eels must live out part of their lives in the sea, where they undergo a marked metamorphosis. *Anguilla vulgaris* is a catadromous fish, compelled to return to the sea to reproduce. Reproduction is thought to take place at great depths in the Mediterranean and Atlantic, and elvers, the semi-transparent larval form, are known to ascend rivers in great numbers.[158]

3.30 (below) The Egyptian eel, Anguilla vulgaris. Dashur, Dynasty VI, Egyptian Museum, Cairo. See Pl. XI
3.31 (right) Anguilla vulgaris from the Tomb of Mereruka, Saqqara, Dynasty VI.

Egyptian eels are dull yellow during the non-breeding season but turn silver in freshwater during the breeding season before going to the sea. The fish is thought to die after breeding, but this fact has not been substantiated for the Egyptian eel. It does not, however, re-enter freshwater after breeding. Only barren specimens remain in the Nile throughout their lives; they can grow to an enormous size.[159]

Mitchell[160] in his reports on the edible fish of Lake Manzala states that the eel is found in large numbers at the entrance of Lake Manzala during the month of December, when they are migrating to the sea to spawn. In January, according to Mitchell, large numbers of young eels are caught entering the lake.

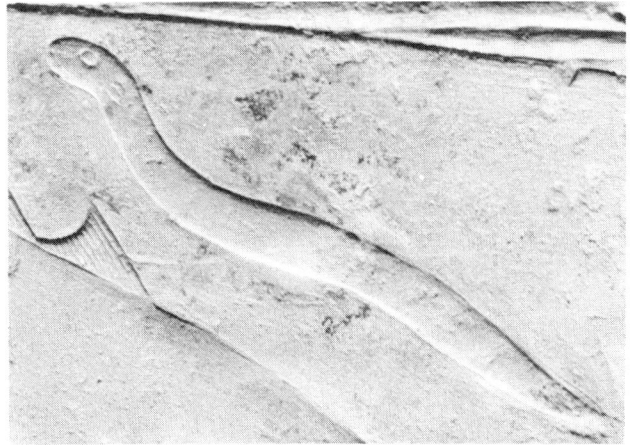

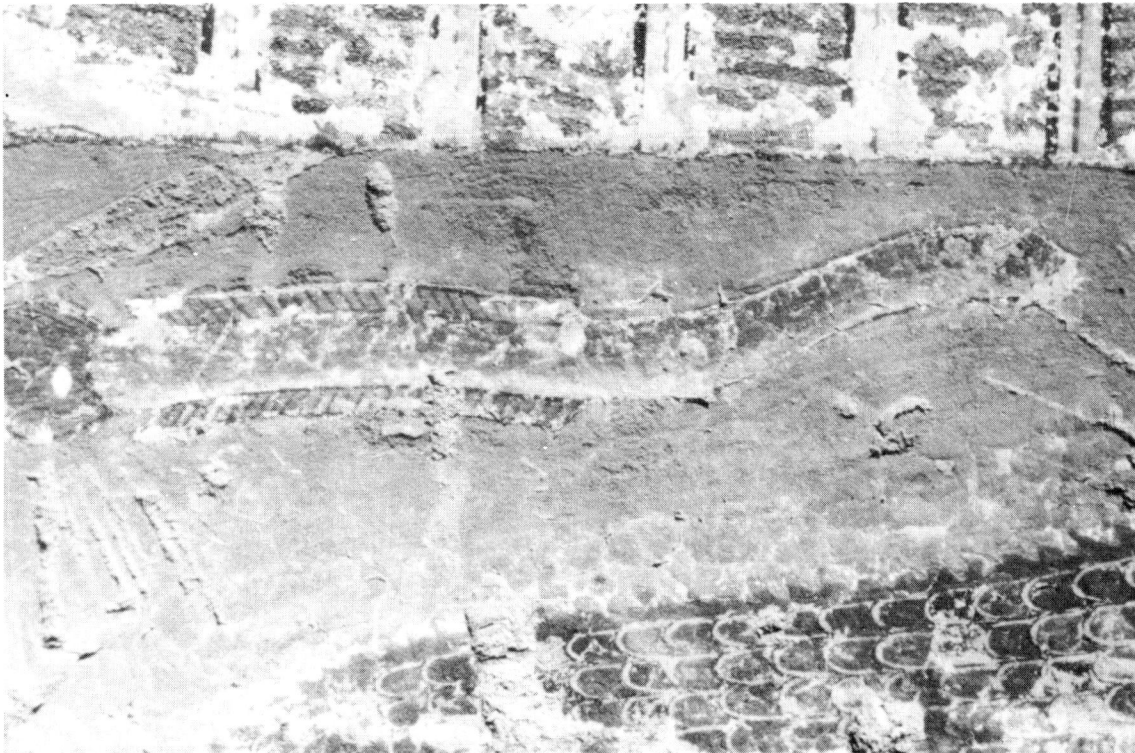

ORDER: MUGILIFORMES

FAMILY: MUGILIDAE (mullets)

This family can be identified by its elongate body and two well separated dorsal fins, the anterior fin composed of hard spines.[161] These fish inhabit the fresh waters and coasts of temperate and tropical regions; the known species number about 100.

Genus: *Mugil*
Egyptian names: *bourie, tobar, garan.*

The genus *Mugil* is represented on the Mediterranean coast of Egypt by three species (*M. cephalus, M. capito*, and *M. auratus*) all of which enter the Nile and have been found far up-river. The mullets are among the most important commercial fish in Egypt today. In commercial catches they probably exceed all other fish, with the exception of *Tilapia (Sarotherodon)*. According to Mitchell[162], the three species of mullet could be identified by the fishermen at Damietta, and all possessed different names: *M. cephalus – bourie, M. capito – tobar*, and *M. auratus – garan.* Investigations conducted for this volume about the Egyptian names for fish show, however, that most Nile river fishermen refer to mullets in general as *bourie*, although they did recognize the existence of different kinds of *bourie*. Pregnant females were sometimes referred to by the name *hut.* Only in areas where mullets were plentiful, such as the Delta lakes, were the fishermen able to give specific names to the different species.

As a family, the mullets are essentially shore fishes, but they have a preference for the mouths of rivers and cutoff lakes where the water is brackish. *Mugil cephalus*, which can be caught near Cairo and is said to range as far south as Aswan, is the *bourie* of the Nile and Delta. It is the most abundant of the three species of mullet found in the Egyptian lakes. *Mugil capito* also readily enters freshwaters and has been recorded as far south as the first cataract. It is thought to prefer shallower waters than *M. cephalus. Mugil capito* is most plentiful in the markets during the summer.[163] In June the fry are seen entering the lakes from the sea. *Mugil auratus* seldom enters rivers and Loat[164] did not obtain any specimens from freshwater. This species is thought to be caught chiefly during the rise and fall of the Nile, when freshwater is mixed with salt water. *Mugil auratus* carries roe from May until August. It is often taken in the same net as *M. capito.*

The grey mullets (*M. cephalus, M capito, M. auratus*) are sociable and move about in large schools, thus rendering their capture easy. They can be found in all parts of Lake Manzala but seem to frequent the deeper water and areas where water plants are abundant. They are remarkable jumpers, able to leap out of the water as much as 1m. and travel more than 3m. before re-entering the water.[165]

All mullets spawn in the sea. The annual migration of the *bourie* (*M. cephalus*) is said to take place from May to November.[166] When about to leave the lake to spawn, males and females congregate in large schools. These schools are made up predominantly of male fish.[167] During this time their natural wariness is diminished and they are easily caught by the fishermen. The roe of the mullet is highly prized. It is often salted, pressed, and dried for sale in the market, where it is known as *batarakh.*

The grey mullet eats decomposing organic matter and feeds on sewage from nearby villages. Despite these eating habits, the mullet is highly esteemed as food by the Egyptian populace.

On the basis of location of the dorsal fin, two forms of mullets have been identified in the fishing scenes at Saqqara. *Mugil cephalus* is thought to be represented by the form showing separated dorsal fins and *M. capito* is represented by the figure with the dorsal fins touching.[168] Unfortunately, an identification based on this characteristic cannot be substantiated by the zoological literature. Although it is possible that the ancient artists were indeed trying to represent two different types of mullet (figs. 3.32, 3.33), it is difficult to assess which artistic form represents which fish. The ichthyological literature places both dorsal fins on both taxa, relative to other landmarks on the fish, in exactly the same place. Consequently, there is no zoological basis for assigning a species epithet to any of the figures. The overall shape of the fish, length of the pectoral fin, and presence to a greater or lesser degree of an adipose eyelid are given as the significant zoological characteristics.[169] Therefore, all mugiliform fish are referred to simply as *Mugil* sp.

3.32 (top) Mugil sp. (upper fish) from the Tomb of Idout. Saqqara, Dynasty V. (The figure below is identified as cf. Gnathonemus cyprinioides)

3.33 (bottom) An example of Mugil sp. Tomb of Ti, Saqqara, Dynasty V.

ORDER: PERCOMORPHI

SUBORDER: PERCOIDEA (perch-like fishes)
FAMILY: CENTROPOMIDAE

Most members of the family Centropomidae are marine. One species, *Lates niloticus*, is found in the freshwaters of Egypt. Although referred to in the English vernacular as the Nile perch, true perch of the family Percidae are confined to the freshwater temperate regions of Europe, Asia, and North America. The family Centropomidae is, however, closely related to Percidae, and members of both families were at one time included in the family Serranidae (sea perches). An important diagnostic characteristic of the family Centropomidae is a dorsal fin that is divided into two parts by a deep notch. The anterior part of the dorsal fin is composed of 7 or 8 spines and the posterior part, one spine and 10-14 rays.[170]

Genus: *Lates*
Species: *L. niloticus*
Synonyms: *Perca nilotica, Perca latus, Centropomus niloticus*
Egyptian names: *isher bayad, samoos, laffash.*

Four species are known for this genus but only one, *L. niloticus*, has been identified from the Egyptian Nile and Lake Nasser.[171] This species is easy to identify by its general shape, strongly serrated preoperculum, and an operculum ending in a spine. The dorsal fin is divided into two parts; the anterior part is supported by numerous spines of which the third is the longest. The posterior dorsal fin is composed of spines (usually one but sometimes two) and soft rays.[172]

Lates niloticus has a wide distribution; it has been recovered throughout the Nile and from most East African lakes. This fish grows to a tremendous size and examples of specimens more than 70kg. have been recorded from all major river systems. An exceptionally large specimen recovered from Lake Nasser measured 2m. in length and weighed 175 kg.[173] The Nile perch is a fierce fighter when caught by hook and line, making it a popular sport fish. To catch *Lates*, successful sport fishermen use a long line with live bait, mainly *Tilapia nilotica* (*Sarotherodon niloticus*) or *Labeo* spp. Commercially, *Lates* is generally caught by trammel-nets.[174] There is always a good market for the fish because it is highly esteemed for its flavor. *Lates* prefer deep, well oxygenated waters.[175] This fish is intolerant of even moderately deoxygenated waters and dies of asphyxiation if the oxygen content of the water falls below a level that would not seriously affect most other species.[176]

At the beginning of this century Lake Qarun was famous for its large Nile Perch. According to the fishermen of the day, far more *Lates* were caught in the winter than in the summer because in winter the fish come near the surface or into shallow water, but in summer they keep to the cooler deep water, which cannot be easily worked with a net.

Loat[177] states that *L. niloticus* was far more numerous in Lakes Manzala and Qarun during the high Nile. He further states that during his three-year survey he only witnessed two fish of an unusually large size. Relatively speaking, Loat's statement holds true today; large specimens are not abundant, but an inspection at any of Egypt's larger fish markets will reveal that an ample number of large *Lates* (1m. or larger) still exist.

The breeding habits of *L. niloticus* from the Egyptian Nile are not well known. In the Nozha-Hydrodrome, *L. niloticus* are known to spawn in late May[178]; *Lates* living in Lake Nasser are known to possess ripe ovaries in April.[179] The spawning season for *L. niloticus* must be quite long because females with mature eggs have also been recorded in February and during the spring and summer months as well.

Kenchington[180] observed that the gonads of *L. niloticus* from the Blue Nile ripen between December and April; the first spent fish appeared in January; the remaining spent individuals rise to peak proportions in April. He gives warmer water conditions as the reason for the early spawning. Nevertheless, temperature may not be the governing factor because *L. mariae* from Lake Tanganykia, where temperatures do not vary more than 2°C during the entire year, spawn seasonally as well.

Hopson[181] has compiled considerable data concerning the breeding habits of this taxon in Lake Chad. Observations on gonads indicate that spawning could take place at all times of the year, but evidence for spawning in Lake Chad from December through February is lacking. Spawning activity was at its peak in April and May.

Feeding habits of *Lates* change with age. The fry feed on underwater insects; juveniles, on small fish, eggs, and shrimp. Large fish subsist mainly on other fish. *Lates* from Gebel Aulia were found to feed primarily on *Alestes, Hydrocynus, Labeo, Tilapia* (*Sarotherodon*), *Synodontis, Mormyrus,* and *Eutropius. Alestes* was the most abundant fish found in *Lates* stomach contents; *Tilapia* (*Sarotherodon*) was the second most frequent food item; *Synodontis* and *Labeo* were found to be the most infrequent food fish. Pekkola[182] noted that although *L. niloticus* predominantly feeds on fish, molluscs were also recorded in the diet. Sandon and Tayib[183] quote Worthington as saying that *Lates* eat plankton and prawns until they reach a size of about 3cm. long, then *Lates* prey solely on fish. In a more recent

3.34 (top) Lates niloticus from the tomb of Seankhuptah, Saqqara, Dynasty VI.
3.35 (bottom) An excellent depiction of the diagnostic spinous and rayed fins of Lates niloticus.
Tomb of Mereruka, Saqqara, Dynasty VI.

study by Hopson[184], *L. niloticus* from Lake Chad reportedly ate fish in the colder months of the year, prawns and other crustaceans during the warmer months, and molluscs from August to October. Latif and Khallaf[185] maintain that immature *Lates* in Lake Nasser stay nearer to shallow waters than adults and eat fresh water shrimp, insect larvae, and cichlid eggs. As the *Lates* mature, they move to deep water, preferring areas with rocky or irregular bottoms, and feed predominantly on fish. Preference toward shallow water by young *Lates* may be a defensive measure because the adults feed on younger, smaller *Lates*. Studies on habitat preference in Lake Nasser show that areas with rocky or irregular bottoms have the highest *Lates* production; those areas with sandy bottoms having no sheltered areas provide the lowest *Lates* production.[186] *Lates*, when mature (ca. 40cm. TL), prefer rock crevices or other secure places and only dart out of these sanctuaries to attack prey[187] and, one must assume, to breed.

Lates has been frequently found mummified. Its likeness has been identified on numerous tombs and is second only to *Tilapia (Sarotherodon)*, and *Mugil* in the frequency of portrayal (fig. 3.34, 3.35). Greek coins from Latopolis (Esna) also displayed the likeness of *Lates*, after which the town was named.

75

FAMILY: CICHLIDAE

Members of the family Cichlidae have been recorded from South America, Africa, and Asia. In Africa, cichlids are the predominant percomorph fish; about 210 species have been identified living in African waters.[188] Cichlids have even been identified from Pleistocene rocks found around Lake Victoria.[189] One reason for the large number of species is that in each major lake or river system a unique flock of species exhibiting a wide range of adaptive types have evolved during geologically recent times.[190]

Three genera are known from the Egyptian Nile: *Hemichromis*, *Haplochromis*, and *Tilapia* (*Sarotherodon*). *Sarotherodon* has recently been designated a distinct taxon and includes two economically important fish formerly included under the genus *Tilapia*: *S. niloticus* and *S. galilaea*. The revised nomenclature has, however, not been applied to the *Tilapia* of Egypt in a consistent matter. Additionally, a controversy exists over the changing of species epithets (e.g., *Tilapia nilotica* to *Sarotherodon niloticus*). Consequently, because the *Tilapia* fishes of Egypt are so closely related, and because the recent literature from Egypt about Nile fish still retains the name *Tilapia* over *Sarotherodon*, the established nomenclature for these fish has been retained throughout this volume.

Most members of this family are mouth-brooders; this characteristic is one of the main features used to separate *Sarotherodon* from *Tilapia*. Mouth brooders shelter their eggs and young in their mouths, which during the breeding season are enlarged for this purpose.[191] For most species the female carries the fertilized eggs and the young larvae in her mouth. Even when the young are free-swimming, the female guards the school and takes the brood back into her mouth at the first sign of approaching danger. Those species that do not practice mouth brooding guard their nest and young for some time after spawning.[192]

Although the cichlids do not achieve the same large size as *Lates*, the fisheries of many African countries are based on the exploitation of *Tilapia*.[193] *Tilapia* provides about 70% of the total fish production of Egypt.[194]

3.36 *Tilapia and Lates, speared by the bident. Tomb of Mereruka, Saqqara, Dynasty VI.*

Genus: *Tilapia*

Synonyms: *Coptodon, Haligenes, Chromis, Oreochromis, Sarotherodon*

Egyptian name: *bolti*

Seventy species of *Tilapia* have been identified from North Africa and the Middle East.[195] Eight species have been described from the Nile[196], but only three, *T. nilotica*, *T. galilaea*, and *T. zillii*, are recovered with any frequency from the Egyptian Nile. Separating these three species can be quite difficult. Breeding habits, skull morphology, tooth-row characteristics, gill-raker counts, and to some extent coloration are used to separate the taxa. However, color varies considerably within each species and an individual can change color during its lifetime.

A reasonably good field characteristic that can be used to separate *T. galilaea* from *T. nilotica* is the presence of vertical markings on the rounded caudal fin of *T. nilotica* and the absence of any markings on the truncated caudal fin of *T. galilaea*. Additionally, *T. nilotica* usually has oblique alternating rows of light and dark spots on its dorsal and anal fin. When breeding, *Tilapia nilotica* displays bright red flanks and a shiny black ventral side. Blue augments the lower gill area.[197] When adult, *Tilapia galilaea* is usually void of any clear markings.[198] Yellow dots on the tail of *T. zillii* separate it from *T. galilaea* and *T. nilotica*. Although they are not always distinct, *T. zillii* possesses 6 to 8 vertical dark bars on the side of its body.[199] These taxa, however, are best separated on the basis of scale and gill-rakers counts. Nevertheless, a trained eye can usually make a good field identification on the basis of color and subtle differences in the shape of the head and nape.

Tilapia nilotica is the most common member of this family in the Nile. *Tilapia galilaea* is encountered much less frequently than *T. nilotica*, and *T. zillii* is the least frequently encountered common species of the *Tilapia* group. *Tilapia* spp. are known to inhabit inshore waters, especially sheltered bays. They are particularly fond of shallow-water areas where water vegetation is abundant.[200]

Tilapia is one of the most frequently portrayed fish in ancient Egypt. It also is one of the oldest identified members of the superclass Pisces represented in Egypt. The oldest representations are in the form of slate palettes, and early pottery effigies recovered from predynastic graves also appear to be that of a *Tilapia*.[201] *Tilapia* is also featured prominently in fishing scenes where a bident has been used to capture fish. In the majority of these cases the *Tilapia* is found speared on one point of the bident and a *Lates* speared on the other point (fig. 3.36). This is an interesting combination because

3.37 *Tilapia sp. from the Tomb of Mereruka. Saqqara, Dynasty VI.*

77

Pl. XIII, fig. 3.38 Tilapia sp. Dashur, Dynasty VI, Egyptian Museum, Cairo.

it is unlikely that such large examples of these taxa would be close enough to each other to be caught with the same thrust. Interestingly, the best fishing grounds for *Lates* would be in Upper Egypt; while *Tilapia* would be most abundant in the swampy shallow water areas of the Delta. By showing a characteristic product of the Upper Nile, *Lates*, next to a characteristic product of the Lower Nile, *Tilapia*, perhaps the artists were attempting to portray a common theme in ancient Egyptian art – the unity of the two lands. The pairing of *Lates* and *Tilapia* on a bident spear is a persistent theme in Old Kingdom and Middle Kingdom tombs and does extend into the New Kingdom although less frequently.[202] In many cases, New Kingdom tombs will display two *Tilapia* speared on the bident rather than a *Tilapia* and a *Lates*.

The food of *T. nilotica* and *T. galilaea* seems to consist of periphytes algae and weeds, whereby the convoluted long intestine acquires a greenish color due to the plant materials.[203] Observations made on stomach contents show, however, that much of the plant material remains undigested.[204] It appears that the fish may depend on the diatoms growing on the plants rather than the plants themselves.[205] In waters where plankton is scarce, this fish eats insects and is known to consume crustaceans as well.[206]

In preparation for spawning, nests are prepared in shallow waters on sandy or sandy-loamy bottoms of near-shore areas. These nests may be of considerable size. Nests more than 1m. in diameter and greater than 40cm. deep can be detected along the shore when the water recedes. In some areas where conditions for nest building are optimum, more than 100 nests are built per hectare.

After the eggs are laid, they are taken up by the female and stored in the bucal and pharyngeal cavities until they have hatched. Even after the young have hatched, they will return to the parent's mouth when frightened. After the period of parental care has terminated, the young usually live in calm waters and frequently move in groups of varying numbers close to the shore.[207]

In Lake Nasser, the ovaries of *T. nilotica* contain two to seven generations of yolky eggs. Females possess ripe yolky eggs (more than 2 mm in diameter) throughout most of the year. (Fully ripe eggs average about 3 mm in diameter.) Consequently, spawning can occur during any season. In other parts of Egypt, the spawning season is shorter. In the vicinity of Cairo, spawning begins in April and extends unti September. In Lake Mariut spawning starts in April, reaches a maximum in May and June, decreases in July and August, and ceases by September.[208]

Tilapia zillii, unlike *T. nilotica* and *T. galilaea*, feeds largely on the leaves and stems of rooted aquatic plants and their associated epiphytic algae.[209] Because of this trait, *T. zillii* has become an important fish for pond cultures.

Tilapia zillii belongs to a groups of cichlids that are not mouth-brooders. Spawning takes place in a prepared nest similar to those of *Tilapia nilotica* and *T. galilaea* – a depression formed on the substrate. After the eggs have been deposited, one of the parents remains on guard, constantly cleaning the eggs of any detritus that may settle on them by maintaining a current of water over the nest.

Boulenger[210] states that the fishermen at Lake Manzala distinguish *T. zillii* and *T. nilotica* by different names. Recent research in this area has shown, however, that the word *bolti* is used to cover all species of *Tilapia*. When asked, the fishermen recognize that different kinds of *bolti* existed in the Nile but explain that they are all called *bolti*.[211]

The Egyptian portrayal of *Tilapia* limits the identification of this fish to the generic level (figs 3.37, 3.38). It seems likely that the ancients were aware of different kinds of *bolti*, as evidence from the New Kingdom pond scenes show (fig. 3.39). However, the zoological characteristics used to identify any single taxon can usually be found in any number of unlikely combinations. For example, the striped tail of *T. nilotica* is shown with a body displaying dark vertical bars, a characteristic more typical of adult *T. zillii*.

3.39 *A pond scene showing several types (species ?) of Tilapia. Nebamun's Garden, Thebes, Dynasty XVIII. British Museum, London. See Pl. XII.*

79

ORDER: TETRAODONTIFORMES
FAMILY: TETRAODONTIDAE (puffer fish)

Genus: *Tetraodon*
Species: *T. fahaka*
Synonyms: *Tetrodon, Crayracion*
Egyptian names: *tambera, fahaka*

The family Tetraodontidae as well as the genus *Tetraodon* is predominantly composed of marine fishes. Four species of the genus *Tetraodon* are freshwater; one species is found in Egypt – *Tetraodon fahaka*. The species epithet for this genus is taken from the Egyptian vernacular name, *fahaka*.

Tetraodon is a puffer fish and has the ability to inflate itself with air. Its likeness can be easily recognized in the tomb scenes at Saqqara by its globe-like body, single dorsal fin, and small pectoral fins (fig. 3.40, 3.41). The name *Tetraodon* refers to its four beak-like teeth, which are extremely powerful and are used for crushing mollusc shells.[212] Even specimens that are nearly dead can easily bite through the end of a finger. According to the Nile fishermen, the *fahaka* rises to the surface to inflate its body; this habit has been verified by St. Hilaire.[213] Inflation is a useful defense for the fish, whereby it increases the breadth of its body making it more difficult to

be devoured by predacious fishes. Nile fishermen are able to make the fish inflate by rubbing its belly or by blowing in the fish's mouth like a balloon. On occasion, usually as a means to startle unsuspecting foreigners, the fishermen will strike an inflated fish with a large stone or hammer; the sound that is emitted resembles that of a small-caliber firearm.

Tetraodon fahaka is usually caught in shallow waters about 5m. deep and is said to prefer to live on sandy or muddy bottoms.[214] It subsists on molluscs and bottom animals, although grasses have also been recovered from the intestines of this fish.[215] Members of this genus are usually poisonous, and *Tetraodon fahaka* is not an exception.[216] The toxin is called *Tetraodon* acid or tetraodontoxin and is classified as a gastro-toxin. In members of this family the poison is usually found in the liver or gonads, but can also be found in the head, gut, or bones of some species. The poison, when artificially concentrated, is a white powder, soluble in water, but insoluble in organic solvents. A lethal dose for humans is 4 mg/kg body weight. No antidote exists for *Tetraodon* poisoning, so treatment is largely symptomatic.[217] Interestingly, the Dinkas of the Sudan consider this fish edible.[218]

3.40 (opposite) *Tetraodon fahaka is easily identified by its bloated, globe-like appearance. Tomb of Mereruka, Saqqara, Dynasty VI.*
3.41 (right) *See frontispiece for this beautiful portrayal of Tetraodon fahaka in color. Tomb of Nebamun, Thebes, Dynasty XVIII British Museum, London.*

81

APPENDIX I
OSTEOLOGY

Most faunal specialists working in Egypt will agree that there is a deficiency of published material illustrating the osteology of Egyptian vertebrates. This deficiency is most obvious in the area of piscine skeletal remains. The varied nomenclature regarding the different elements of the fish skeleton adds further confusion. The purpose of this appendix is to provide an illustrated guide to the diagnostic skeletal elements of the Nile fish most commonly encountered in Egyptian archaeological sites. The illustrations are not intended to serve as a substitute for comparative materials, but rather as a supplement to a comparative collection. Unfortunately, such collections are generally not available in most of our museums and even when present are not often loaned due to the delicate nature of the skeletons. Building a basic comparative collection is, therefore, an important first step in the identification of fish remains. Although this is a difficult and time-consuming endeavor, it is the only avenue open to the archaeologist interested in the identification of piscine remains.

Scholars interested in building a comparative collection are referred to Boulenger (1907), Greenwood (1966), Sandon (1950) and Latif (1974) for assistance in identifying Nile fish. For a synonomy of the fish skeleton, interested readers should refer to Starks (1901).

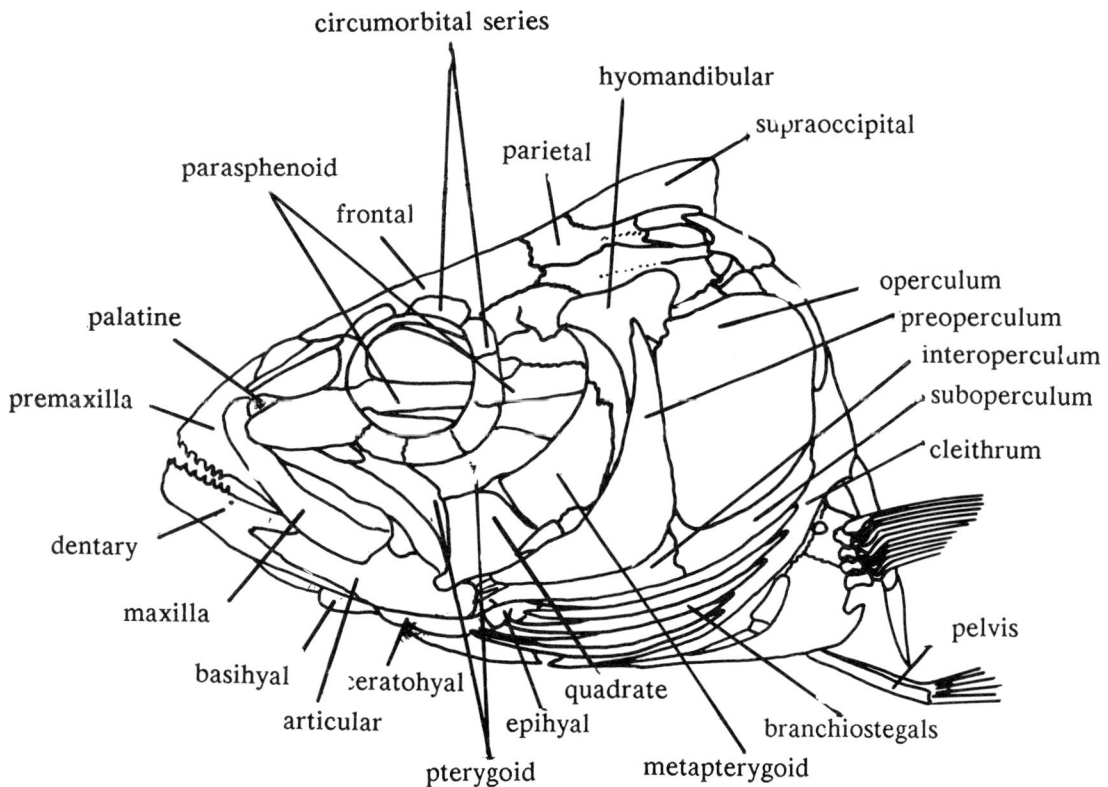

Fig. A–1 The Syncranium. A composite diagram based on the identifiable elements most frequently recovered from Egyptian archaeological sites (after Gregory 1933, fig. 7).

83

Fig A-2
LATES NILOTICUS

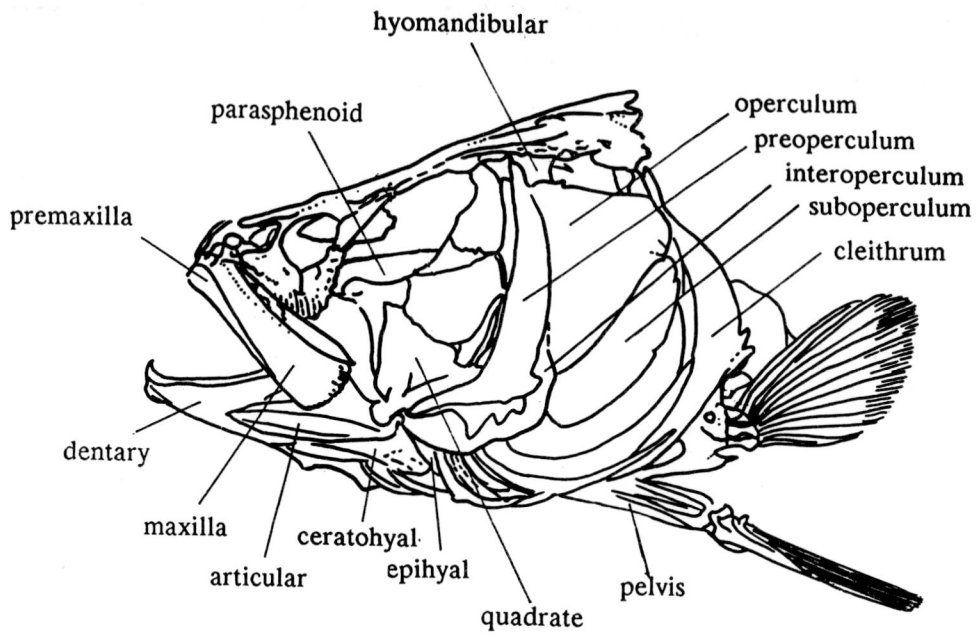

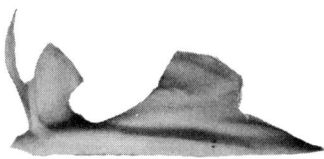

premaxilla

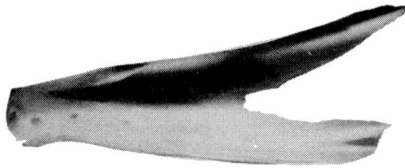

dentary

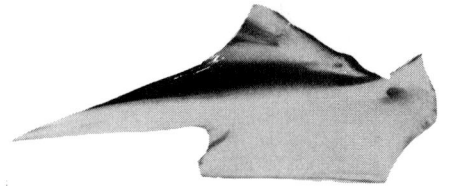

articular

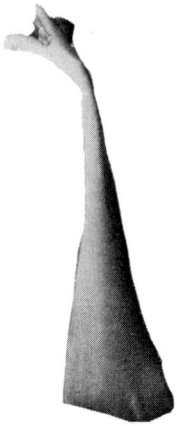

maxilla

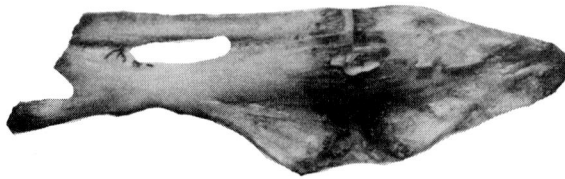

ceratohyal-epihyal

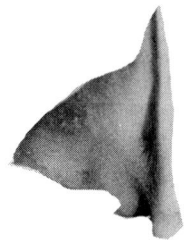

quadrate

pelvis

84

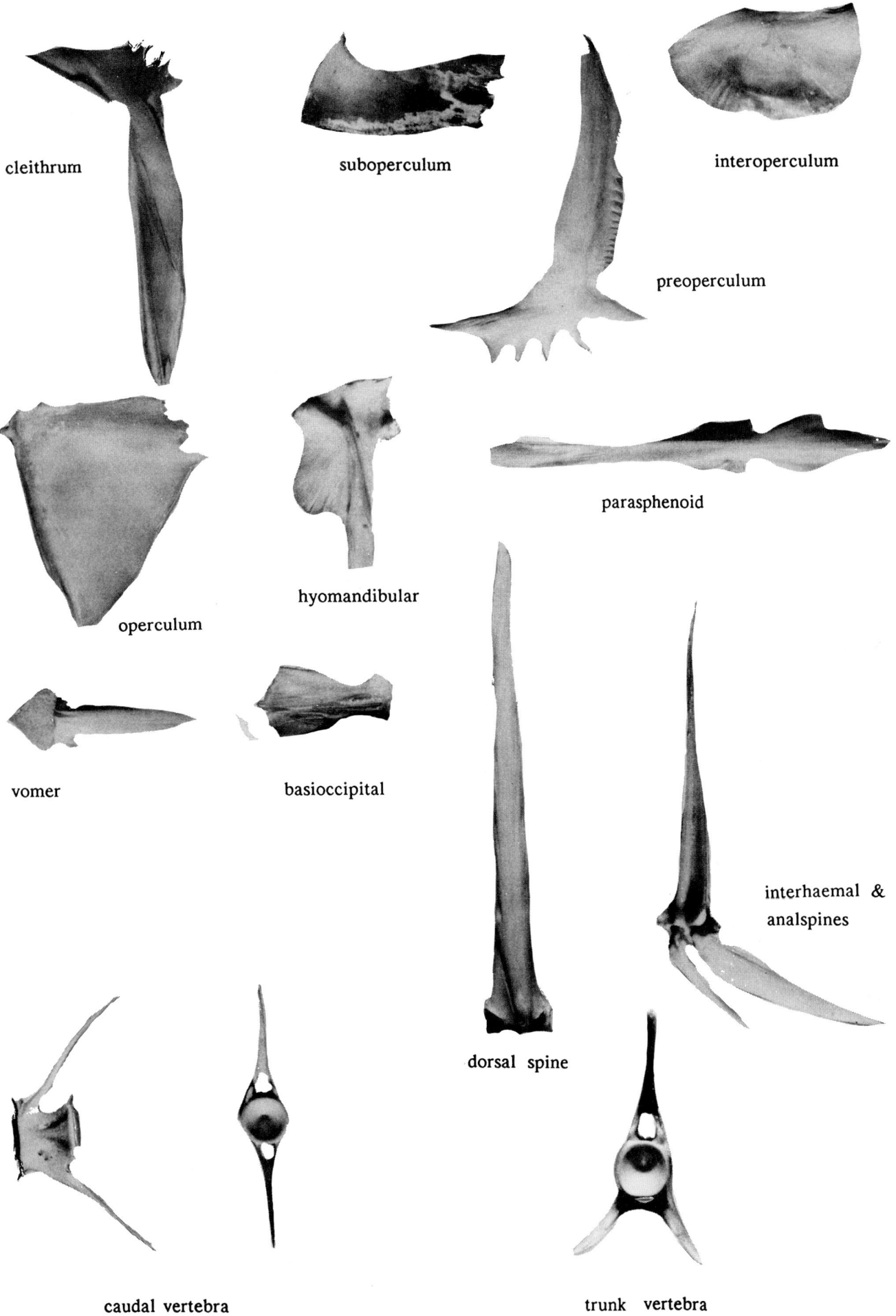

cleithrum

suboperculum

interoperculum

preoperculum

operculum

hyomandibular

parasphenoid

vomer

basioccipital

interhaemal &
analspines

dorsal spine

caudal vertebra

trunk vertebra

85

Fig A–3
CLARIAS SP.

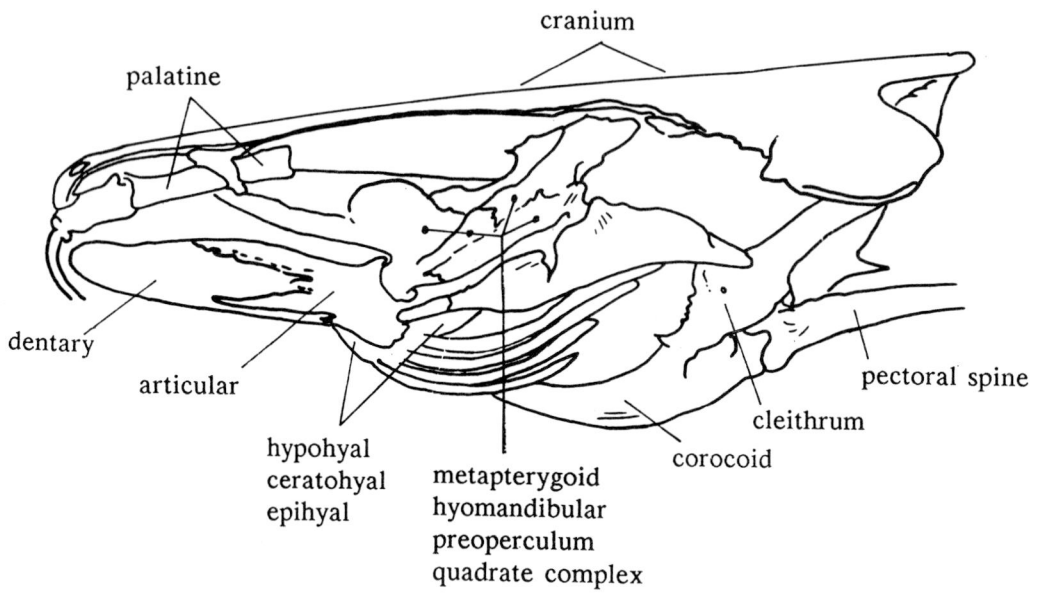

cranium

palatine

dentary

articular

hypohyal
ceratohyal
epihyal

metapterygoid
hyomandibular
preoperculum
quadrate complex

corocoid

cleithrum

pectoral spine

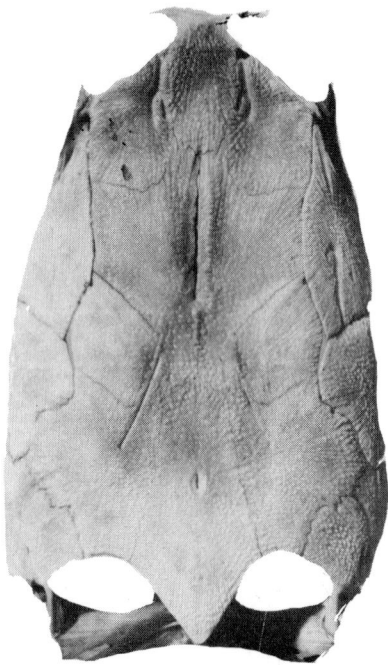

cranium

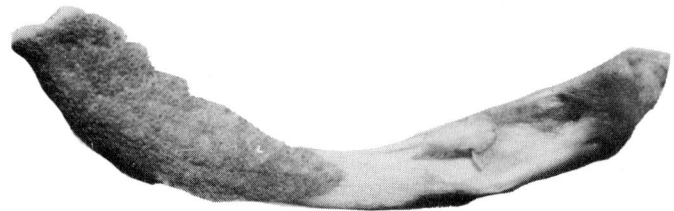

dentary & articular

metapterygoid
hyomandibular
preoperculum
quadrate complex

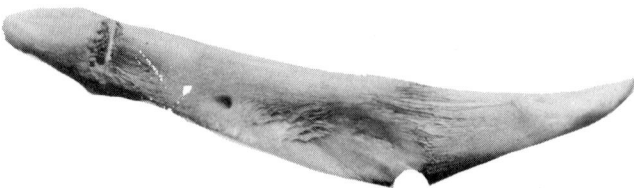

hypohyal – ceratohyal – epihyal

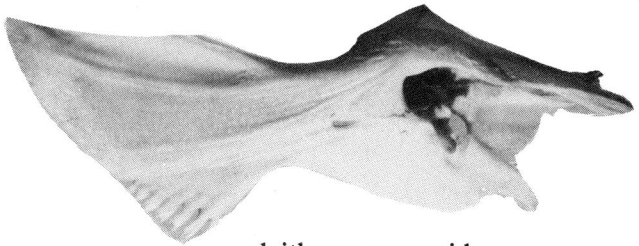

cleithrum-corocoid

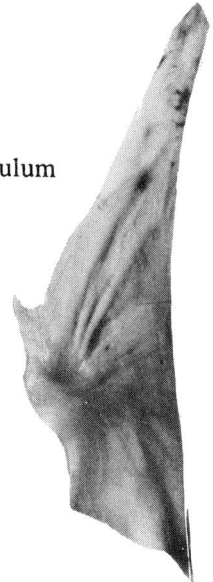

operculum

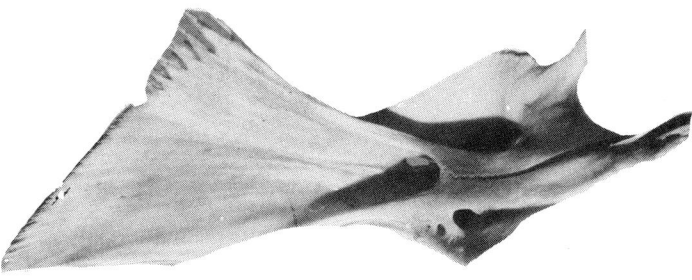

cleithrum-corocoid

urohyal

palatine

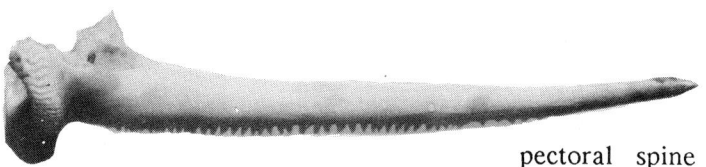

pectoral spine

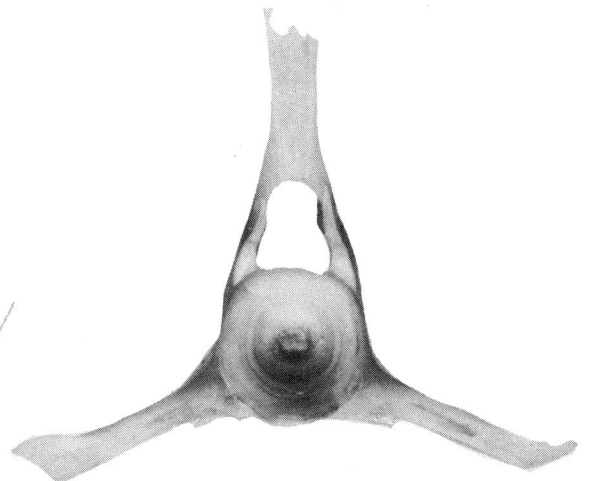

trunk vertebra

caudal vertebra

Fig A–4
TILAPIA SP.

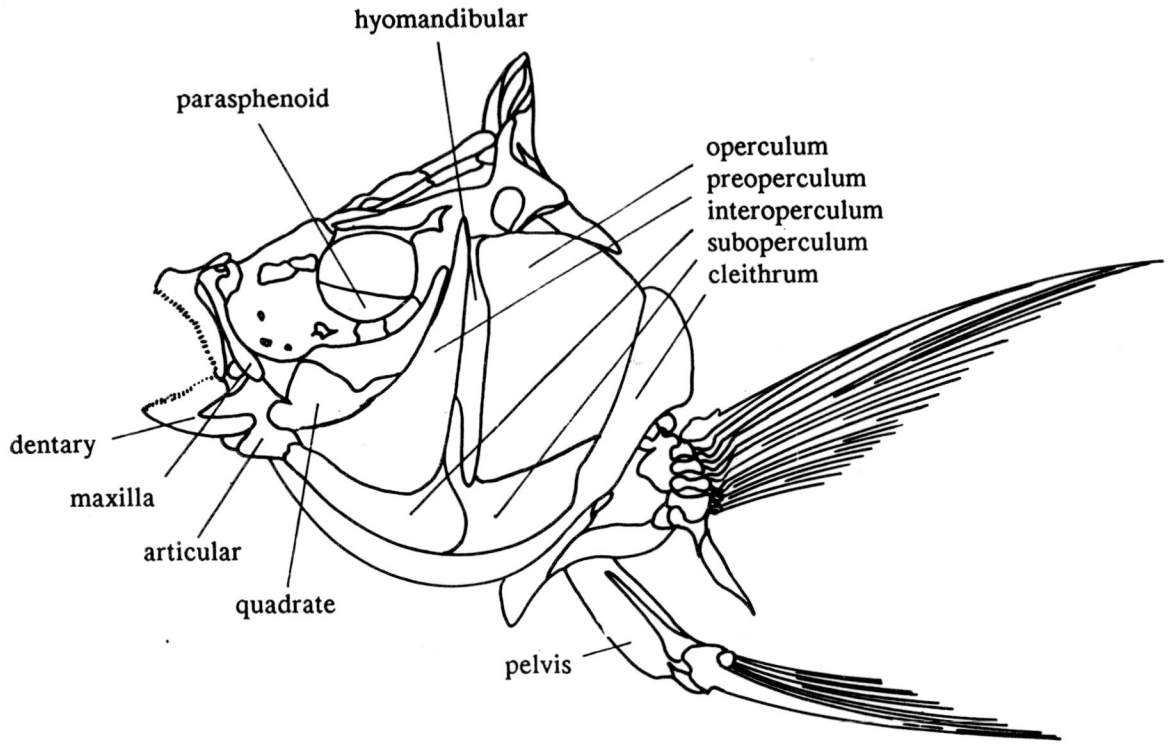

parasphenoid

hyomandibular

operculum
preoperculum
interoperculum
suboperculum
cleithrum

dentary

maxilla

articular

quadrate

pelvis

premaxilla

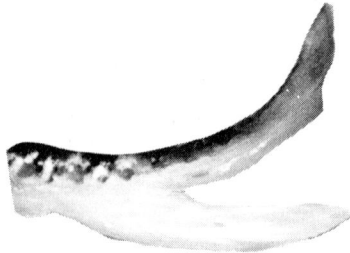

dentary

articulaı

maxilla

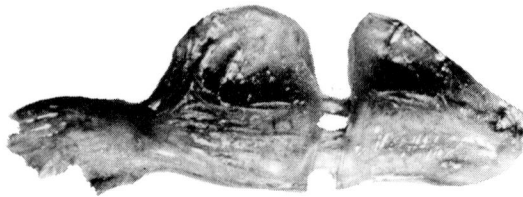

ceratohyal-epihyal

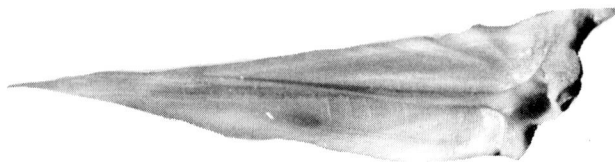

pelvis

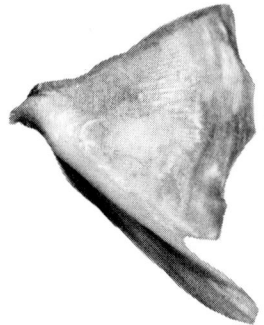

quadrate

88

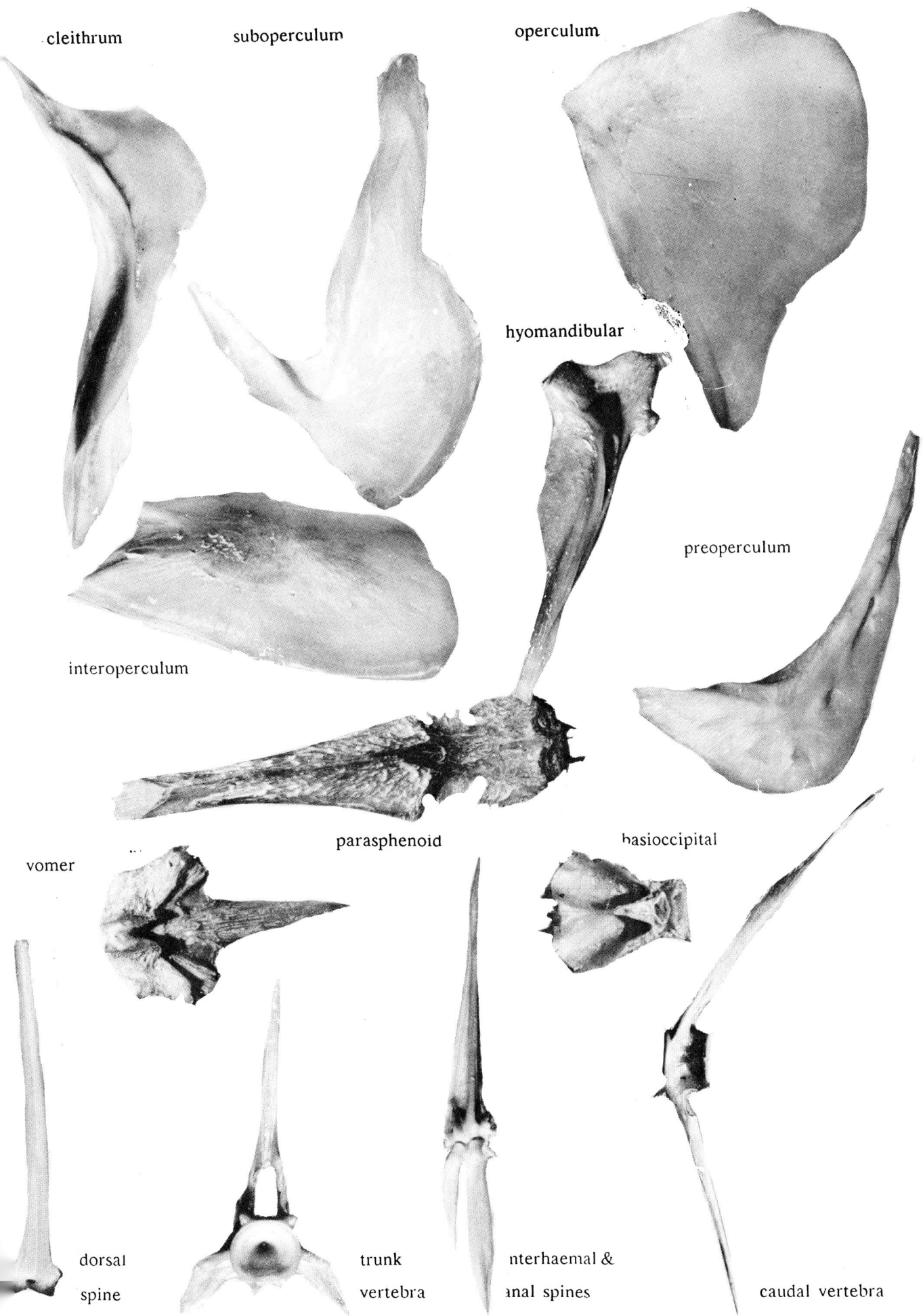

cleithrum

suboperculum

operculum

hyomandibular

preoperculum

interoperculum

vomer

parasphenoid

basioccipital

dorsal
spine

trunk
vertebra

nterhaemal &
anal spines

caudal vertebra

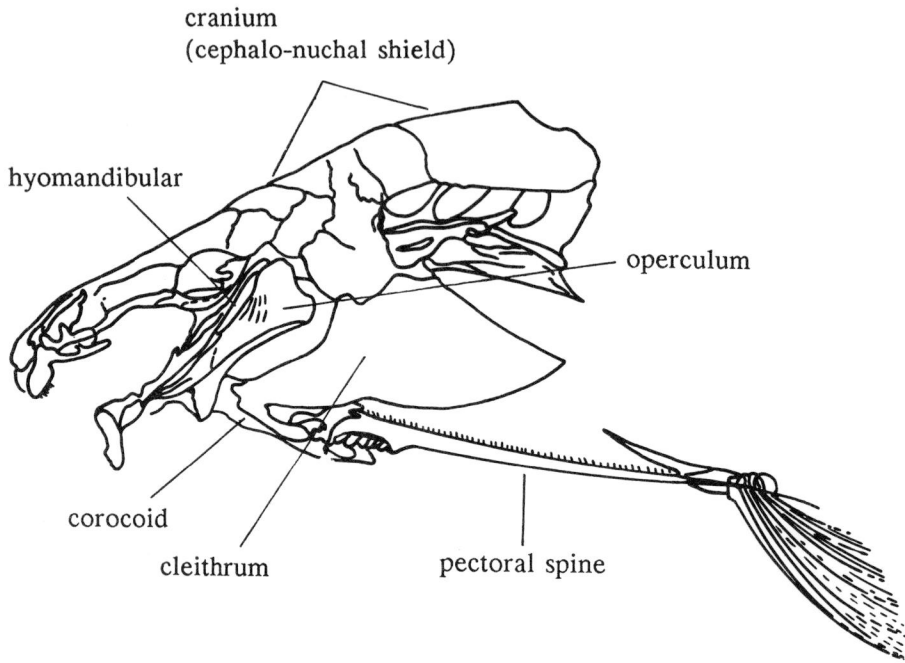

cranium
(cephalo-nuchal shield)

hyomandibular

operculum

corocoid

cleithrum

pectoral spine

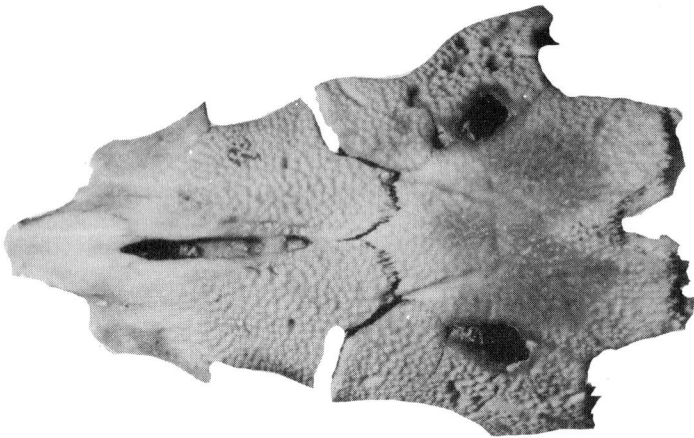

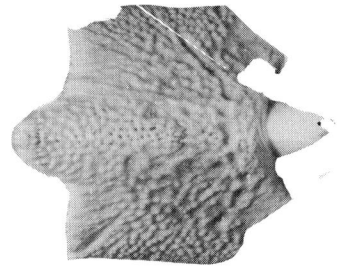

cephalo-nuchal shield & 1st modified dorsal spine

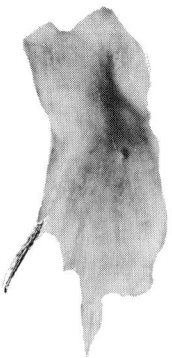

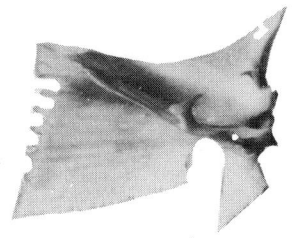

hyomandibular

epihyal-ceratohyal

corocoid

cleithrum

operculum

pectoral spine

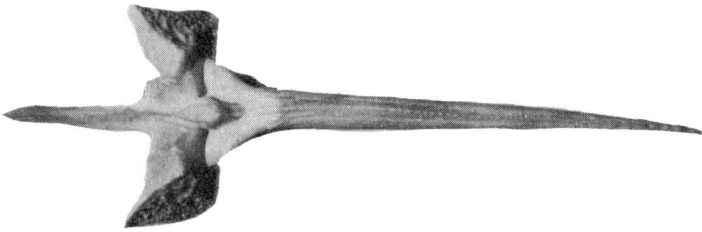

pterygiophore (modified) & 2nd dorsal spine

1st modified vertebra

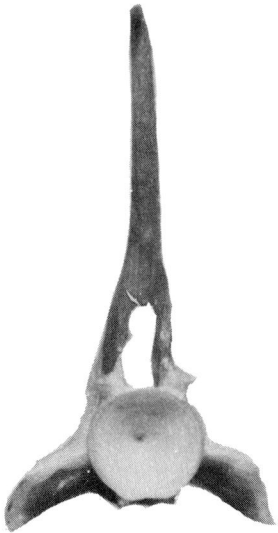

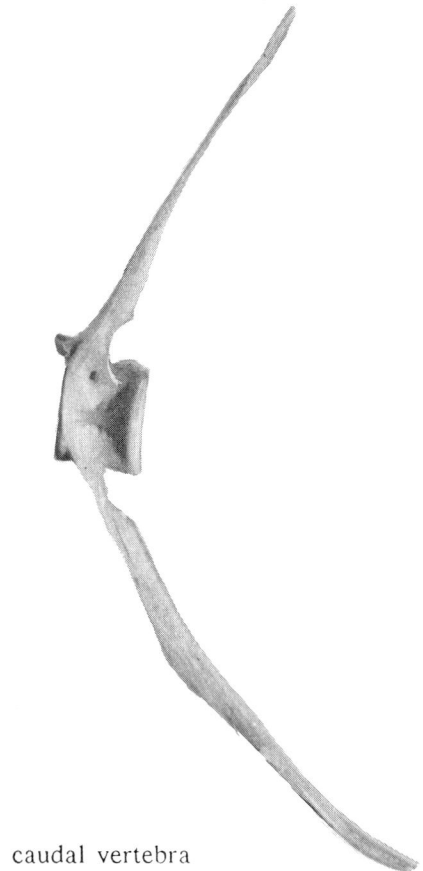

trunk vertebra

caudal vertebra

BIBLIOGRAPHY

ABU-GIDEIRI 1984. Y. B. Abu-Gideiri. *Fishes of the Sudan*. Khartoum, 1984.

AELIAN 1832. Aelian. *De Natura Animalium*. X. F. Jacobs ed., 1832.

ALTENMÜLLER 1973. H. Altenmüller. "Bemerkungen zum Hirtenlied des Alten Reiches," *Chronique d'Égypte* 48, No. 96 (1973), pp. 211-231.

ANDERSON 1902. J. Anderson. *Zoology of Egypt: Mammalia*. London, 1902.

ARKELL 1949. A. Arkell. *Early Khartoum*. Oxford, 1949.

BATES 1917. O. Bates. "Ancient Egyptian Fishing," *Harvard African Studies* 1 (1917), pp. 199-272.

BAUMGARTEL 1960. E. J. Baumgartel. *The Cultures of Prehistoric Egypt II*, Oxford, 1960.

BIDOLI 1976. D. Bidoli. *Die Sprüche der Fangnetze Alägyptischen Sargtexten*, Abhandlungen des Deutschen Archäologischen Instituts Kairo 9, 1976.

BISHAI and ABU-GIDEIRI 1965. H. M. Bishai and Y. B. Abu-Gideiri. "Studies of the Biology of Genus *Synodontis* at Khartoum: Food and Feeding Habits". *Hydrobiologia Acta Hydrobiologica Hydrographica et Protistologica* XXVI-II, (1965), pp. 98-113.

BISHAI and ABU-GIDEIRI 1967. H.M. Bishai and Y.B. Abu-Gideiri. "Studies on the biology of Genus *Synodontis* at Khartoum: Classification and Distribution" in *Revue de Zoologie et de Botanique Africaines*. LXXV, (1967), pp. 1-2, 17-30.

VON BISSING 1905, 1911. F. W. von Bissing. *Die Mastaba des Gem-ni-kai I & II*. Berlin, 1905, 1911.

BLACKMAN 1914-1953. A. M. Blackman & M. R. Apted. *The Rock Tombs of Meir*. Vols I-VI, London, 1914-1953.

BOESSNECK and VON DEN DRIESCH 1982. J. Boessneck and A. Von den Driesch. *Studien an subfossilen Tierknochen aus Ägypten*, Münchner Ägyptologische Studien 40, Munich, 1982.

BORCHARDT 1913. L. Borchardt. *Das Grabdenkmäl des König, Sahure I, II*. Leipzig, 1910-1913.

BOULENGER 1907. G. Boulenger. *The Fishes of the Nile*. London, 1907.

BOURRIAU 1981. J. Bourriau. *Umm el Ga'ab Pottery from the Nile Valley before the Arab Conquest*. Cambridge, 1981.

BOURRIAU 1988. J. Bourriau. *Pharaohs and Mortals: Egyptian Art in the Middle Kingdom*. Cambridge, 1988.

BREASTED 1906-7. J. H. Breasted. *Ancient Records of Egypt*, 5 vols. Chicago 1906-7

BREWER 1986. D. J. Brewer. *Cultural and Environmental Change in the Faiyum Egypt: An Investigation Based on Faunal Remains*. Ph.D Dissertation. University of Tennessee, University Microfilms, Ann Arbor, 1986.

BREWER 1987a. D. J. Brewer. "Seasonality in the Prehistoric Faiyum Based on the Incremental Growth Structures of the Nile Catfish (Pisces: *Clarias*)". *Journal of Archaeological Science* 14, (1987), pp. 459-472.

BREWER 1987b. D. J. Brewer. "A Report on the Aquatic Fauna from HK29A,". *A Final Report to the National Endowment for the Humanitites on Predynastic Research at Hierakonpolis, 1985-86*, M. Hoffman (ed), University of South Carolina, 1987, pp. 45-47.

BROOKS 1950. J. L. Brooks. "Speciation in Ancient Lakes," *Quarterly Review of Biology*, 25 (1950), pp. 131-176.

BRUNTON 1927. G. Brunton. *Qau and Badari I*. London, 1927.

BRUNTON 1937. G. Brunton. *Mostagedda and the Tasian Culture*. London, 1937.

BRUNTON and CATON-THOMPSON 1928. G. Brunton and G. Caton-Thompson. *The Badarian Civilization and Predynastic Remains near Badari*. London, 1928.

BRUTON 1979. M. N. Bruton. "The Breeding Biology and Early Development of *Clarias gariepinus* (Pisces: Clariidae)," *Transactions of the Zoological Society of London* 35 (1979), pp. 1-45.

BUDGE 1967. E.A.W. Budge. *The Egyptian Book of the Dead*. New York, 1967.

BUTZER 1976. K. Butzer. *Early Hydraulic Civilization in Egypt*. Chicago, 1976.

CAMINOS 1956. R. A. Caminos. *Literary Fragments in the Hieratic Script*. Oxford, 1956.

CAPART 1939. J. Capart. "Cultes d'El Kab et Préhistoire," *Chronique d'Égypte* 14, No. 28 (1939), pp. 213-217.

CATON-THOMPSON and GARDNER 1934. G. Caton-Thompson and E. W. Gardner. *The Desert Fayum*. London, 1934.

ČERVÍČEK 1974. P. Červíček. *Felsbilder des Nord-Etbai, Oberägyptens und Unternubiens*. Wiesbaden, 1974.

CHRISTOPHE 1967. L-A. Christophe. "Le ravitaillement en poissons des artisans de la nécropole Thébaine à la fin du règne de Ramsès III". *Bulletin de l'Institut Français d'Archéologie Orientale*, Cairo 65 (1967), pp. 176-199.

CORBET 1961. P.S. Corbet. "The Food of Non-cichlid Fishes in the Lake Victoria Basin, with Remarks on Their Evolution and Adaptation to Lacustrine Conditions". *Proceedings Zoological Society London* 136 (1961), pp. 1–101.

DAGET 1954. J. Daget. "Les poissons du Niger supérieur," *Mémoirs Institut Français Africaine noire* 36 (1954), pp. 1–391.

DAGNAN-GINTER et al., 1984. A. Dagnan-Ginter, B. Ginter, J. Kozlowski, M. Pawlikowski, and J. Sliwa. "Excavations in the Region of Qasr el-Sagha, 1981. Contributions to the Neolithic Period, Middle Kingdom Settlement and Chronological Sequences in the Northern Fayum Desert." *Mitteilungen des Deutschen Archäologischen Instituts Abteilung Kairo*, 40 (1984), pp. 33–102.

DAMBACH and WALLERT 1966. M. Dambach and I. Wallert. "Das *Tilapia*-Motiv in der altägyptischen Kunst," *Chronique d'Égypte* 41, No. 82 (1966), pp. 273–294.

DANELIUS and STEINITZ 1967. E. Danelius and H. Steinitz. "The Fishes and Other Aquatic Animals on the Punt Reliefs at Deir El-Bahri," *Journal of Egyptian Archaeology* 53 (1967), pp. 15–24.

DARBY et al., 1977. W.J. Darby, P. Ghalioungui, and L. Grivetti. *Food: the Gift of Osiris I, II.* London, 1977.

DAUMAS 1964. F. Daumas. "Quelques remarques sur les représentations de pêche à la ligne sous l'Ancien Empire," *Bulletin de l'Institut Français d'Archéologie Orientale*, Cairo 62 (1964), pp. 67–85.

DAUMAS 1975. F. Daumas. "Fischer und Fischerei," *Lexikon der Ägyptologie* 2 (1975), pp. 234–242.

DAVIES 1901. N. de G. Davies. *The Rock Tombs of Sheikh Said*, London, 1901.

DAVIES 1902. N. de G. Davies. *The Rock Tombs of Deir el Gebrâwi* I, II, London, 1901, 1902.

DAVIES 1903. N. de G. Davies. *The Rock Tombs of El Amarna Part I. The Tomb of Meryra*, London, 1903.

DAVIES 1905. N. de G. Davies. *The Rock Tombs of el Amarna, Part 3. The Tombs of Huya and Ahmes.* London, 1905.

DEBONO 1948. F. Debono. El-Omari (prés de Hélouan), Exposé sommaire sur les campagnes des fouilles 1943–1944 et 1948. *Annales du Service des Antiquités de l'Égypte* 48 (1948), pp. 561–569.

DEBONO and MORTENSEN (in press). F. Debono and B. Mortensen. *El-Omari: A Neolithic Settlement and Other Sites in the Vicinity of Wadi Hof, Helwan.* Archaelogische veröffentlichungen, Deutschen Archäologischen Instituts, Abteilung Kairo, in press.

DESROCHES-NOBLECOURT 1954. C. Desroches-Noblecourt. "Poissons, tabous et transformations du mort," *Kêmi* 13 (1954), pp. 33–42.

DESROCHES-NOBLECOURT 1966. C. Desroches-Noblecourt. "Une fiole evoquant le poisson "*Lates*" de la déesse Neith," *Mélanges offerts à Kazimiers Michalowski.* Warsaw, 1966, pp. 71–81.

DIODORUS SICULUS 1933. *Library of History.* C.H. Oldfather (transl.). Vol 1, Cambridge, Mass.

VON DEN DRIESCH 1983. A. von den Driesch. *Some Archaeological Remarks on Fishes in Ancient Egypt, Animals and Archaeology: 2. Shell Middens, Fishes, and Birds.* British Archeological Reviews, International Series 183 (1983), pp. 87–110.

VON DEN DRIESCH 1986. A. von den Driesch. "Tierknochenfunde aus Qasr el Sagha/Fayum," *Mitteilungen des Deutschen Archäologischen Instituts, Abteilung Kairo*, 42 (1986), pp. 1–8.

DUELL 1938. P. Duell. *The Mastaba of Mereruka I, II.* Chicago, 1938.

DUNHAM 1935. D. Dunham. "A Palimpsest on an Egyptian Mastaba Wall". *American Journal of Archaeology* 39 (1935), pp. 300–309.

EDEL 1961. E. Edel. *Zu den Inschriften auf den Jahreszeitenreliefs der "Weltkammer" aus den Sonnenheiligtum des Niussere I.* Nachrichten der Akademie der Wissenschaften zu Göttingen (phil. hist. *Kl.*) 8 (1961), pp. 209–255.

EDEL 1963. E. Edel. *Zu den Inscriften auf den Jahreszeitenreliefs der "Weltkammer" aus den Sonnenheiligtum des Niussere II.* Nachrichten der Akademie der Wissenschaften zu Göttingen (phil. hist. *Kl.*) 4/5 (1963), pp. 89–217.

EDEL 1976. E. Edel. *"Der Tetrodon fahaka als Bringer der Überschwemmung und sein Kult im Elephantengau"* Mitteilungen des Deutschen Archäologischen Instituts, Abteilung Kairo, 32 (1976), pp. 35–43.

EDEL 1978. E. Edel. "Noch einmal zum Kult des *Tetrodon fahaka* als Bringer der Überschwemmung im Elephantengau," *Göttinger Miszellen* 30 (1978), pp. 35–37.

EDEL 1980. E. Edel. "*Sṯpw* "Springer" als Bezeichnung der Mugiliden. Der älteste Beleg für die Anlage von Mugildenteichen zur Vorratshaltung". *Orientalia* 49 (1980), pp. 204–207.

EIWANGER 1979. J. Eiwanger. "Zweiter Vorbericht über die Wiederaufnahme der Grabungen in der neolithischen Siedlung Merimde-Benisalâme," *Mitteilungen des Deutschen Archäologischen Instituts, Abteilung Kairo* 35, (1979), pp. 23–57.

EIWANGER 1984. J. Eiwanger. *Merimde-Benisalâme I: Die Funde der Urschicht,* Archäologische Veröffentlichungen, Deutsches Archäologisches Insituts, Abteilung Kairo, 47 1984.

ELSTER and JENSEN 1960. H.J. Elston and K.

W. Jensen. "Limnological and Fishery Investigations of the Nozha-Hydrodrome near Alexandria, Egypt 1954–1956," *Alexandria Institute of Hydrobiology, Notes and Memoirs* 43 (1960), pp. 1–99.

EMERY 1938. W. B. Emery. *The Tomb of Hemaka*. Cairo, 1938.

EMERY 1962. W. B. Emery. *A Funerary Repast in an Egyptian Tomb of the Archaic Period*. Leiden, 1962.

EPHRON and DAUMAS 1939. L. Ephron, F. Daumas and G. Goyon. *Le Tombeau de Ti*. Mémoires de l'Institut Français d'Archéologie Orientale du Caire, 65 fasc 1, Cairo, 1939.

ERMAN 1918. A. Erman. *Reden Rufen und Lieder auf den Graberbildern des Alten Reich*. Berlin, 1918.

ERMAN and GRAPOW 1926. A. Erman and H. Grapow. *Wörterbuch der Ägyptischen Sprache* I–VIII. Berlin, 1926.

FARAG 1982. S. Farag. "Une inscription memphite de la XIIe dynastie". *Revue d'Égyptologie* 32 (1982), pp.74–84.

FISCHER 1977. H.G . Fischer. "Some Iconographic and Literary Comparisons," *Fragen an die altägyptische Literatur. Studien zum gedenken an Eberhard Otto*. J. Assman, E. Feucht, and R. Grieshammer (eds). Weisbaden 1977, pp. 155–170.

FISCHER 1983. H. G. Fischer. *Ancient Egyptian Calligraphy*. Metropolitan Museum of Art. New York, 1983.

FISH 1956. G.R. Fish. "Some Aspects of the Respiration of Six Species of Fish from Uganda," *Journal of Experimental Biology* 33 (1956), pp. 186–195.

GAILLARD 1923. M. C. Gaillard. *Recherches sur les poissons représentés dans quelques tombeaux Égyptiens de l'Ancien Empire*, Mémoires de l'Institut Français d'Archéologie Orientale du Caire 51, 1923, pp. i–136.

GAMER-WALLERT 1970. I. Gamer-Wallert. *Fische und Fischkulte im alten Ägypten*. Ägyptologische Abhandlungen 21, Wiesbaden, 1970.

GAMER-WALLERT 1975a. I. Gamer-Wallert. "Fische, profan". *Lexikon der Ägyptologie* Vol II (1975), pp. 224–228.

GAMER-WALLERT 1975b. I. Gamer-Wallert. "Fische, religios,". *Lexikon der Ägyptologie* Vol II (1975), pp. 228–234.

GARDINER 1976. A. Gardiner. *Egyptian Grammar*. 3rd edition. Oxford, 1976.

GAUTIER et al., 1980. A. Gautier, P. Ballmann, and W. Van Neer. "Molluscs, Fish, Birds, and Mammals from the Late Palaeolithic Sites in Wadi Kubbaniya", *Loaves and Fishes: the Prehistory of Wadi Kubbaniya*, F. Wendorf and R. Schild (eds), 1980, pp. 281–284.

GLANVILLE 1926. S. R. K. Glanville. "A Note on Herodotus II.93," *Journal of Egyptian Archaeology* 12 (1926), pp. 75–76.

GODRON 1949. G. Godron. "À propos du nom royal "Nr-Mr", *Annales du Service des Antiquités de l'Égypte,* Cairo 49 (1949), pp. 217–219.

GREENWOOD 1951. P.H. Greenwood. "Evolution of the African Cichlid Fishes, the *Haplochromis* Species Flock in Lake Victoria" *Nature* 167 (1951), pp. 19–20.

GREENWOOD 1955. P.H. Greenwood. "Reproduction in the Catfish *Clarias mossambicus* Peters," *Nature* 176 (1955), pp. 516–518.

GREENWOOD 1956. P.H. Greenwood. "A Revision of the Lake Victoria *Haplochromis* Species (Pisces, Cichlidae) Part I," *Bulletin Museum of Natural History, Zoology* 4 (5) (1956), pp. 223–244.

GREENWOOD 1963. P.H. Greenwood. "A Collection of Fish from the Aswa River Drainage System Uganda," *Proceedings Zoological Society of London* 140 (1963), pp. 61–74.

GREENWOOD 1965. P.H. Greenwood. "Explosive Speciation in African Lakes," *Proceedings Royal Institute* 40, (184) (1965), pp. 256–269.

GREENWOOD 1966. P.H. Greenwood. *The Fishes of Uganda*. Kampala, 1966.

GREENWOOD 1968. "Fish Remains". In *The Prehistory of Nubia I*. F. Wendorf (ed.), Fort Burgwin Research Center and Southern Methodist University pp. 100–109, 1968.

GREENWOOD 1974. P.H. Greenwood. "The *Haplochromis* Species (Pisces: Cichlidae) of Lake Rudolf, East Africa," *Bulletin British Museum of Natural History* (Zoology) 27 (1974), pp. 139–165.

GREENWOOD and TODD 1970. P.H. Greenwood and E. J. Todd. "Fish Remains from Olduvai," *Fossil Vertebrates from Africa*. Vol II, L.S.B. Leaky and R.J.G. Savage (eds), pp. 225–241, London, 1970.

GREENWOOD and TODD 1976. P. H. Greenwood and E. J. Todd. "Fish remains from Upper-Paleolithic sites near Idfu and Isna". *Prehistory of the Nile Valley*, F. Wendorf and R. Schild (eds), pp. 383–388. New York, 1976.

GREGORY 1933. W. K. Gregory. Fish skulls: A study of the evolution of natural mechanisms. *Transactions of the American Philosophical Society*, vol xxiii (1933).

GRIFFITH and NEWBERRY 1895. F. Griffith and P. Newberry. *El-Bersheh II*. London, 1895.

GUGLIELMI 1973. W. Guglielmi. "*Reden, Rufen, und Lieder auf altägyptischen Darstellungen der Landwirtschaft, Viehzucht, des Fisch- und Vogelfangs vom Mittleren Reich bis zur Spätzeit*. Tübinger Ägyptologische Beiträge 1, 1973.

HARPUR 1987. Y. Harpur. *Decoration in the*

Tombs of the Old Kingdom. London, 1987.

HASHEM 1973. M.T. Hashem. "The Age, Growth, and Maturity of *Labeo niloticus*, Forsk. from the Nozha-Hydrodrome in 1969–1970," *Bulletin of the Institute of Oceanography and Fisheries*, 3. (1973), pp. 85–94.

HANDOUSSA 1988. T. Handoussa, "Fish offering in the Old Kingdom", Mitteilungen des Deutschen Archäologischen Instituts, Abteilung Kairo 44, (1988), pp. 105–109.

HASHEM and EL-AGAMY 1977. M.T. Hashem and A. El-Agamy. "Effects of Fishing and Maturation on the *Barbus bynni* Population of the Nozha-Hydrodrome," *Bulletin of the Institute of Oceanography and Fisheries* 7 (1977), pp. 363–392.

HASHEM and HUSSEIN 1973. M.T. Hashem and K.A. Hussein. "Some Biological Studies of the Nile Perch, *Lates niloticus*, in the Nozha-Hydrodrome," *Bulletin of the Institute of Oceanography and Fisheries* 3 (1973), pp. 363–382.

HASSAN 1986. F. Hassan. "Holocene Lakes and Prehistoric Settlements of the Western Faiyum, Egypt," *Journal of Archaeological Science* 13 (1986), pp. 483–501.

HASSAN 1938. S. Hassan. "Excavations at Saqqara 1937–1938," *Annales du Service des Antiquités de l'Égypte, Cairo* 38 (1938), pp. 503–521.

HAYES 1953. W.C. Hayes. *The Scepter of Egypt Part I*. New York, 1953.

HELCK 1954. W. Helck. "Herkunft und Deutung einiger Züge des frühägyptischen Königsbildes," *Anthropos* 49 (1954), pp. 961–991.

HELCK 1965. W. Helck. *Materialien zur Wirtschaftsgeschichte des Neuen Reiches*. Part 5. Weisbaden, 1965.

HENEIN 1988. N. H. Henein. *Mari Girgis. Village de Haute-Égypte*. Bibliothèque d'Étude 94, Institut Français d'Archéologie Orientale du Caire. Cairo, 1988.

HERODOTUS 1926. *Historis*, II, A.D. Godley (transl.). Loeb Classical Library, 117. Cambridge Mass., 1926.

HOBSON 1987. C. Hobson. *Exploring the World of the Pharoahs*. London, 1987.

HOFFMAN 1982. M. A. Hoffman. *The Predynastic of Hierakonpolis*, Egyptian Studies Association I, Cairo, 1982.

HOPSON 1972. A. J. Hopson. *A Study of the Nile Perch in Lake Chad*. Foreign and Commonwealth Office Overseas Development Research Publication 9, 1972.

HOPSON 1982. A.J. Hopson. *Lake Turkana. A Report on the Findings of the Lake Turkana Project 1972–1975*. Government of Kenya and The Ministry of Overseas Development Vol. 1, 1982, London.

HORNUNG 1982. E. Hornung. *Conceptions of god in Ancient Egypt*. Ithaca 1982.

VAN DER HORST 1982. P. van der Horst. "The Way of Life of the Egyptian Priests According to Chaeremon," *Studies in Egyptian Religion Dedicated to Professor Jan Zandee*. M. Van Voss (ed), pp. 61–71, Leiden, 1982.

JANSSEN 1961. J. M. A. Janssen. *Two Ancient Egyptian Ships' Logs*. Leiden, 1961.

JANSSEN 1975. J. J. Janssen. *Commodity Prices from the Ramessid Period*. Leiden, 1975.

JAROS-DECKERT 1984. B. Jaros-Deckert. *Grabung im Asasif 1963–1970. Band V: Das Grab des Jnj-jtj.f. Die Wandmalereien der XI, Dynastie*. Archäologische Veröffentlichung 12, Deutsches Archäologisches Institut, Abteilung Kairo,

JUNKER 1929. H. Junker. "Vorläufiger Bericht uber die Grabung auf der neolithischen Siedlung von Merimde-Benisalame". *Anzeiger der Akademie der Wissenshaft in Wein. (Phil. Hist. Kl.)* xviii: (1929), pp. 156–250.

KANAWATI 1980–1985. N. Kanawati. *The Rock Tombs of El Hawawish. The Cemetery of Akhmim*. I–V. Sydney, 1980–5.

KAPLONY 1963. P. Kaplony *Die Inschriften der ägyptischen Frühzeit*, Ägyptologische Abhandlungen 8, Weisbaden, 1963.

KEIMER 1939. L. Keimer "La boutargue dans l'Égypte ancienne," *Bulletin de l'Institut d'Égypte* 21, Cairo (1939), pp. 215–243.

KEIMER 1948. L. Keimer. "Quelques représentations rares de poissons égyptiens remontant à l'époque pharaonique," *Bulletin de l' Institut d'Égypte* 29, Cairo (1948), pp. 263–274.

KESSLER 1985. D. Kessler. "Tierkult," *Lexikon der Ägyptologie* VI (1985), pp. 571–587.

KENCHINGTON 1939. F. E. Kenchington. "Observations on the Nile Perch (*Lates niloticus*) in the Sudan," *Proceedings Zoological Society of London 101* (1939), pp. 157–168.

KLEBS 1915. L. Kleb. *Die Reliefs des alten Reich*. Heidelberg, 1915.

KROMER 1978. K. Kromer. *Siedlungsfunde aus dem frühen Alten Reich in Giseh*. Wien, 1978.

KRONIG 1934. W. Kronig. "Ägyptische Fayence Schalen des Neuen Reiches," *Mitteilungen des Deutschen Archäologischen Instituts, Abteilung Kairo* 5 (1934), pp. 144–167.

KRZYZANIAK 1977. L. Krzyzaniak. *Early Farming Cultures on the Lower Nile*. Travaux du Centre d'Archéologie Méditerranéenne de l'Académie Polonaise des Sciences 21, 1977.

LACAU 1913. P. Lacau. "Suppressions et modifications de signes dans les textes funéraires" *Zeitschrift für Ägyptische Sprache und Altertumskunde* 51 (1913), pp. 1–64.

LACAU 1954. P. Lacau. "Le panier de pêche égyptien". *Bulletin de l'Institut Français* d'Archéologie Orientale, Cairo 54, (1984), pp. 136–163.

LAGERCRANTZ 1953. S. Lagercrantz. "Forbidden Fish," *Orientalia Suecana* 2 (1953), pp. 1–8.

LARSEN 1960. H. Larsen. "Knochengeräte aus Merimde in der ägyptischen Abteilung des Mittelmeer museums" *Orientalia Suecana* 9. (1960), pp. 28–53.

LATIF 1974. A.F.A. Latif. *Fisheries of Lake Nasser*. Aswan Regional Planning, Lake Nasser Development Center, Aswan, 1974.

LATIF and KHALLAF 1974. A.F.A. Latif and E. A. Khallaf. "Studies of the Nile Perch, *Lates niloticus*, from Lake Nasser," *Bulletin of the Institute of Oceanography and Fisheries* 4 (1974), pp. 131–164.

LATIF and EL-SAYED 1974. A. F. A. Latif and A. K. el-Sayed. "Studies of the Nile Perch *Lates niloticus* from Lake Nasser". *Bulletin of the Institute of Oceanography and Fisheries* 4 (1974), pp. 131–164.

LEOPOLD 1964. L. Leopold, M. Wolman, J. Miller. *Fluvial Processes in Geomorphology*. San Francisco, 1964.

LEPSIUS 1849-59. C. Lepsius. *Denkmäler aus Aegypten und Aethiopien*. I-XII. Berlin, 1849–1859.

LICHTHEIM 1975. M. Lichtheim. *Ancient Egyptian Literature. A Book of Readings I: The Old and Middle Kingdoms*. Berkeley, 1975.

LICHTHEIM 1976. M. Lichtheim. *Ancient Egyptian Literature. A Book of Readings II: The New Kingdom*. Berkeley, 1976.

LICHTHEIM 1980. M. Lichtheim. *Ancient Egyptian Literature. A Book of Readings III: The Late Period*. Berkeley, 1980.

LLOYD 1975. A. B. Lloyd. *Herodotus Book II, Introduction*. Leiden, 1975.

LLOYD 1982. A.B. Lloyd. "Nationalist Propaganda in Ptolemaic Egypt," *Historia* 32 (1982), pp.33–55.

LOAT 1904. L. Loat. "Gurob" in *Saqqara Mastabas I and Gurob* by M. A. Murray and L. Loat, London, 1904.

LOAT 1907. L. Loat. "Mr. Loat's Report on the Nile Fish Survey," in *The Fishes of the Nile* by G. A. Boulenger (1907), pp. xx–li, London.

LYTHGOE and DUNHAM 1965. A. Lythgoe and D. Dunham. *The Predynastic Cemetery N7000: Naga ed Der*. Berkeley, 1965.

MACRAMALLAH 1935. R. Macramallah. *Le Mastaba d'Idout*. Cairo, 1935.

McARDLE 1982. J. McArdle. "Preliminary Report on the Predynastic fauna of the Hierakonpolis Project". *The Predynastic of Hierakonpolis*. M. Hoffman (ed). The Egyptian Studies Association 1 (1982), pp. 110–115.

McARDLE 1987. J. McArdle. "Preliminary Report on the Mammalian Faunal Remains from HK-29A", *A Final Report to the National Endowment for the Humanities on Predynastic Research at Hierakonpolis 1985-1986*. M. A. Hoffman (ed), Univ. of South Carolina (1987), pp. 42–44.

MEEKS 1973. D. Meeks. "Le nome du dauphin et le poisson de Mendès" *Revue d'Égyptologie* 25, (1973), pp. 209–216.

MENGHIN and AMER 1932. O. Menghin and M. Amer. *Excavations of the Egyptian University in the Neolithic site of Maadi, First Preliminary Report* (season of 1930–31), Cairo, 1932.

MOHR 1943. T. Mohr. *The Mastaba of Hetep-her-ackhti: Study of an Egyptian tomb Chapel in the Museum of Antiquities*. Ex Oriente Lux 5, Leiden 1943.

MOND and MYERS 1937. R. Mond and O. Myers. *Cemeteries of Armant I*. London, 1937.

MONTET 1914. P. Montet. "Les poissons employés dans l'écriture hieroglyphique," *Bulletin de l'Institut Français d'Archéologie Orientale* 11 (1914), pp. 39–48.

MONTET 1925. P. Montet. *Scènes de la vie privée dans les tombeaux égyptiens de l'Ancien Empire*. Strasbourg, 1925.

MONTET 1950. P. Montet. "Le fruit défendu," *Kêmi* 11 (1950), pp. 85–116.

MUOSSA and ALTENMÜLLER 1977. A. M. Moussa and H. Altenmüller. *Das Grab des Nianchchnum und Chnumhotep*. Archäologische Veröffentlichungen, Deutsches Archäologisches Institut, Abteilung Kairo 21. 1977.

MÜLLER 1974. I. Müller. "Die Ausgestaltung der Kultkammern in den Gräbern des Alten Reiches in Giza und Saqqara," *Forschungen und Berichte* 16. Berlin (1974), pp. 79–96.

NEEDLER 1984. W. Needler. *Predynastic and Archaic Egypt in the Brooklyn Museum*. New York, 1984.

VAN NEER 1986. W. Van Neer. "Some Notes on the Fish Remains from Wadi Kubbaniya (Upper Egypt; Late Paleolithic). *Fish and Archaeology*, D. Brinkhuizen and A. Clason (eds). British Archaeological Reviews, International Series 294, (1986), pp. 103–113.

NETOLITZKY 1911. F. Netolitzky. "Nahrungs- und Heilmittel der Urägypter" *Die Umschau* 15 no. 46 (1911), pp. 953–956.

NETOLITZKY 1943. F. Netolitzky. "Nachweise von Nahrungs und Heilmitteln in den Trockenleichen von Naga ed Der (Ägypten). *Mitteilungen des Deutschen Archäologischen Instituts, Abteilung Kairo* Ertes Erganzungsheft, Berlin, (1943). pp. 5–33.

NEWBERRY 1894. P. E. Newberry. *El-Bersheh* I. London 1894.

NEWBERRY and FRASER 1893. P.E. Newberry and G. W. Fraser. *Beni Hasan* I. London, 1893.

PEKKOLA 1919. W. Pekkola. "Notes on Habits, Breeding and Food of Some White Nile Fish". *Sudan Notes and Records* 2 (1919), pp. 112–121.

PETRIE 1890. W. M. F. Petrie. *Kahun, Gurob, and Hawara*. London, 1890.

PETRIE 1892. W. M. F. Petrie. *Medum*. London, 1892.

PETRIE 1900. W. M. F. Petrie. *The Royal Tombs of the First Dynasty I*. London, 1900.

PETRIE 1901a. W. M. F. Petrie. *Diospolis Parva*. London, 1901.

PETRIE 1901b. W. M. F. Petrie. *The Royal Tombs of the First Dynasty at Abydos II*. London, 1901.

PETRIE 1902. W. M. F. Petrie. *Abydos, Part I*. London, 1902.

PETRIE 1915. W. M. F. Petrie. "The Metals in Egypt," *Ancient Egypt I*, (1915), pp. 12–23.

PETRIE 1917. W. M. F. Petrie. *Tools and Weapons*. London, 1917.

PETRIE 1920. W. M. F. Petrie. *Prehistoric Egypt*. London, 1920.

PETRIE 1921. W. M. F. Petrie. *Corpus of Prehistoric Pottery and Palettes*. London, 1921.

PETRIE and QUIBELL 1896. W. M. F. Petrie and J. E. Quibell. *Naqada and Ballas*. London, 1896.

PLUTARCH 1936. Plutarch. "Isis and Osiris" in *Moralia 5*. Translated by F.C. Babbitt. Cambridge Mass, 1936.

PORTER and MOSS 1960. B. Porter and R. Moss. *Topographical Bibliography of Ancient Egyptian Hieroglyphic Texts, Reliefs, and Paintings: I₁, The Theban Necropolis*. Oxford, 1960.

QASIM and QAYYUM 1961. S.Z. Qasim and A. Qayyum. "Spawning Frequency and Breeding Seasons of Some Freshwater Fishes with Special Preference to Those Occurring in the Plains of Northern India," *Indian Journal of Fisheries 8* (1961), pp. 24–43.

QUIBELL 1900. J. E. Quibell. *Hierakonpolis I*. London, 1900.

QUIBELL and GREEN 1902. J. E. Quibell and F. W. Green. *Hierakonpolis II*. London, 1902.

RADCLIFF 1926. W. Radcliff. *Fishing from the Earliest Times*. London, 1926.

RAWLINSON 1866. G. Rawlinson. *Herodotus II*. New York, 1866.

RIEFSTAHL 1972. E. Riefstahl. "A Unique Fish Shaped Glass Vial in the Brooklyn Musuem. *Journal of Glass Studies* 14 (1972), pp. 10–14.

RIZKANA and SEEHER 1988. I. Rizkana and J. Seeher. *Maadi II. The Lithic Industries of the Predynastic Settlement*. Archäologische Veröffentlichungen 65, Deutsches Archäologisches Institut, Abteilung Kairo, 1988.

ROBERTS 1975. T.R. Roberts. "Geographical Distribution of African Freshwater Fishes," *Zoological Journal of the Linnean Society 57* (1975), pp. 249–319.

RUFFER 1919. M. A. Ruffer. "Food in Egypt," *Mémoire Presenté à l'Institut Égyptien 1* (1919), pp. 1–88.

SAAD 1969. Z. Saad. *The Excavations at Helwan: Art and Civilization in the First and Second Egyptian Dynasties*. Oklahoma, 1969.

SANDON 1950. H. Sandon. *An Illustrated Guide to the Freshwater Fishes of the Sudan*. London, 1950.

SANDON and TAYIB 1953. H. Sandon and A. Tayib. "The Food of Some Common Nile Fish," *Sudan Notes and Records* 24 (1953), pp. 205–239.

SÄVE-SODERBERGH 1975. T. Säve-Soderbergh. "Harpune and Harpunieren". *Lexikon der Ägyptologie* Vol II (1975), pp. 1012–1015.

SAVE-SODERBERGH 1953. T. Säve-Soderbergh. *On Egyptian Representations of Hippopotamus Hunting as a Religious Motive*. Horae Söderblominae 3. Uppsala 1953.

SCAMUZZI 1964. E. Scamuzzi. *Egyptian Art in the Egyptian Museum of Turin*. Turin, 1964.

SIMPSON 1976a. W. K. Simpson. *The Offering Chapel of Sekhem-ankh-Ptah in the Museum of Fine Arts*. Boston, 1976.

SIMPSON 1976b. W. K. Simpson. *The Mastabas of Qar and Idu G7101 and G7102*. Boston, 1976.

SLAUGHTER 1968. B. Slaughter. "Additional comments on the Fish Remains". *The Prehistory of Nubia*. F. Wendorf (ed.), Dallas 1968. pp. 933–934.

SMITH 1942. W. S. Smith. "The Origin of some Unidentified Old Kingdom Reliefs," *American Journal of Archaeology* 46 (1942), pp. 509–551.

SMITH 1965. W. S. Smith. *Interconnections in the Ancient Near East*. New Haven, 1965.

STAEHELIN 1978. E. Staehelin. "Zur Hathorsymbolik in der ägyptischen Kleinkunst," *Zeitschrift für ägyptische Sprache und Altertumskunde* 105 (1978), pp. 76–84.

STARKS 1901. E. Starks. "Synonomy of the Fish Skeleton". *Washington Academy of Sciences, Proceedings* 3 (1901), pp. 507–539.

STRABO 1932. *Geography*. H.L. Jones (transl.) VIII. Cambridge Mass. 1932.

STRAUSS 1974. E-C Strauss. *Die Nunschale Eine Gefässgruppe des Neuen Reiches*. Münchner Ägyptologische Studien 30, 1974.

TYLOR 1896. J. Tylor. *The Tomb of Sebeknakht*. London, 1896.

TOUNY and WENIG 1969. A. Touny and S. Wenig. *Sport in Ancient Egypt*. Leipzig, 1969.

VANDIER 1950. J. Vandier. *Mo'alla: La tombe d'Ankhtifi et la Tombe de Sébekhotep*. Bibliothèque d'étude 18, Institut Français d'Archéologie Orientale. Cairo 1950.

VANDIER 1952. J. Vandier. *Manuel d'archaéologie égyptienne I. Les époques de* formation. *Les Trois Premières Dynasties*. Paris 1952.

VANDIER 1964.   J. Vandier.   *Manuel d'archéologie égyptienne IV.   Bas-reliefs et peintures. Scènes de la vie quotidienne Première Partie: Les Tombes.* Paris, 1964

VANDIER 1964b.   J. Vandier.   "Quelques remarques sur la préparation de la boutargue". *Kêmi* 17, pp. 26–34.

VANDIER 1969.   J. Vandier.   *Manuel d'archéologie égyptienne V. Bas-reliefs et peintures. Scènes de la vie quotidienne.* Paris, 1969

VAN DE WALLE 1978.   B. van de Walle. *La chapelle funéraire de Neferirtenef.* Brussels, 1978.

WENDORF and SCHILD 1976.   F. Wendorf and R. Schild. *The Prehistory of the Nile Valley.* New York, 1976.

WENDORF and SCHILD 1980.   F. Wendorf and R. Schild. *Loaves and Fishes: The Prehistory of Wadi Kubbaniya.* Dallas, 1980.

WENKE 1986.   R. Wenke.   "Old Kingdom Community Organization in the Western Delta" *Norwegian Archaeological Review* 19 (1986), pp. 15–33.

WENKE et al., 1983.   R. Wenke, P. Buck, J. Hanley, M. Lane, J. Long, and R. Redding.   "The Fayum Archaeological Project: Preliminary Report of the 1981 Season". *American Research Center in Egypt Newsletter* 122 (1983), pp. 24–34.

WENKE and REDDING 1988.   R. Wenke and R. Redding.   "Excavations at Kom el-Hisn 1986," *Newsletter of the American Research Center in Egypt* 135 (1988), pp. 11–17.

WENKE et al., 1988.   R. Wenke, J. Long, and P. Buck.   "Epipaleolithic and Neolithic Subsistence and Settlement in the Fayyum Oasis of Egypt," *Journal of Field Archaeology* 15 (1988), pp. 29–51.

WENKE and LANE (in press) . R. Wenke and M. Lane. *Land of the Lake.* Malibu, in press.

WHITEHEAD 1959.   P.J.B. Whitehead.   "The Anadromous Fishes of Lake Victoria." *Revue de Zoologie et de Botanie Africaine* 59 (1959) pp. 329–363.

WILKINSON 1883.   G. Wilkinson.   *Manners and Customs of the Ancient Egyptians* 2.   Boston, 1883.

WINKLER 1938.   H.A. Winkler. *Rock Drawings of Southern Upper Egypt I.* London, 1939.

EL-ZARKA Et AL. 1970.   S. El-Zarka, A.H. Shaheen, and A.A. El-Aleem.   "Reproduction of *Tilapia nilotica*," *Bulletin of the Institute of Oceanography and Fisheries* 1 (1970). pp. 193–204.

NOTES

INTRODUCTION

1. Butzer 1976. Leopold, et al. 1964, p. 316.
2. Butzer 1976, pp. 15–18.
3. Hornung 1982, pp. 180–181. Van Neer 1986, p. 106.
4. Van Neer 1986, pp. 107–108. The first weathering experiments were conducted in the Eastern Sahara in 1985.
5. Slaughter 1968, pp. 933–934; Greenwood 1968, pp. 100–109; Gautier et al. 1980, pp. 281–294; Van Neer 1986, one of the researchers involved in the study of the fish remains at the Palaeolithic camps at Wadi Kubbaniya, now agrees with the strong evidence supporting a natural bias against the preservation of certain skeletal elements such as vertebrae.
6. For example see: Helck 1965, pp. 816–828; Caminos 1956.
7. Von den Driesch 1983; Gaillard 1923.
8. Edel 1961, pp. 209–255; Edel 1963, pp. 89–217; Gamer-Wallert 1970, pp. 39–42, 52–53.
9. Gamer-Wallert 1970, pp. 53–54, 124–126; Dambach and Wallert 1966, pp. 273–294.
10. Gamer-Wallert 1970, p. 111; Handoussa 1988, p. 109.
11. The six fish gathered around a similar ball on a small ceramic chest of predynastic date from El Amrah are among the earliest depictions of identifiable fish. Glanville 1926, pp. 75–76. Another predynastic example is a slate palette of two fish joined at the mouth from Nagada tomb 345 (Baumgartel 1960, p. 88). The motif appears again in almost identical fashion, once in the Middle Kingdom on a stela from Abydos below the owner spear fishing. (Dambach and Wallert 1966, pp. 275, 278–283). In the New Kingdom it becomes stylized into a neat optical illusion of three symetrically placed fish whose bodies converge into a single triangular head in the center (Kronig 1934, pp. 144–167; Strauss 1974, p. 19). Although the ball of eggs is symbolic of the rising sun (Gamer-Wallert 1970, p. 113) it appears that the Egyptians confused this behavior with that of mouth brooding and a general theme of fertility and rebirth is maintained. See also Chapter III on *Tilapia* breeding habits.
12. Gamer-Wallert 1970, pp. 116–119; Capart 1939, 213–217. The so-called catfish demons on the purchased Predynastic leather fragment are not convincing.
13. Gamer-Wallert 1970, p. 116.
14. Lichtheim 1975, pp. 215–217; Bourriau 1988, pp. 148–149; Handoussa 1988, p. 108; Gamer-Wallert (1970, pp. 77, 121–122) believes the popularity of this fish is due to its association

with the decan stars 10 and 36, which are depicted on contemporary sarcophagi two *Synodontis*. However, a connection Hathor, the goddess of the swamp and fertility whose cult was popular in the Middle Kingdo especially in the provinces from which majority of these amulets have been found, is more likely explanation (Staehelin 1978, 76–84).
15. Hornung 1982, pp. 79–80.
16. Gamer-Wallert 1970, pp. 88–90, 86–1 See also Gamer-Wallert 1975b, pp. 230–232.
17. Gamer-Wallert 1970, pp. 75–85; Mor 1950, pp. 85–116.
18. Herodotus 1975, pp. 49–170.

CHAPTER 1

1. Wendorf and Schild 1980.
2. Wendorf and Schild 1980.
3. Van Neer 1986, pp. 104–106; Bre 1987a, p. 461; see also Chapter III.
4. Slaughter 1968, pp. 933–934.
5. Greenwood and Todd 1976, pp. 383–388
6. Van Neer 1986, pp. 103–113.
7. Hassan 1986, pp. 483–501.
8. Caton-Thompson and Gardner 1934.
9. Hassan 1986; Wendorf and Schild 19 Wenke et al. 1983.
10. Wenke et al. 1988, p. 38.
11. Brewer 1986, pp. 45–71; Wenke et 1988, pp. 39–44.
12. Brewer 1986, pp. 110–124.
13. Wenke et al. 1988, pp. 40–41.
14. Brewer 1987a, pp. 460–461.
15. Brewer 1986, 1987a; See also Cha III.
16. Brewer 1986, 1987a, and references there.
17. Brewer 1986, pp. 72–124, and refere cited there.
18. Anderson (1902) notes that jackals in Fayum caught fish, presumably Nile catfish, f shallow water pools. Loat (1907) states that water levels in canals and pools were low, *Cl* could be caught by men groping with their h in the mud. Greenwood and Todd (1970) reported that fish of this genus can be e collected by hand during the spawning season.
19. Hopson 1972.
20. See Brewer 1987a for more details.
21. Hassan 1986, p. 497.

22. Dagnan–Ginter et al. 1984, pp. 94–96.

23. Von den Driesch 1986, pp. 1–8.

24. Debono and Mortensen, in press.

25. Brunton and Caton-Thompson 1928, p. 33; Brunton 1937, p.31; Tasian village midden.

26. Brunton and Caton-Thompson 1928, p. 33. Brunton 1937, p. 58, 90, pl. xli 68, 69. Brunton 1927, sect. 159. Kromer 1978, p. 83, pl. 35.

27. Brunton and Caton-Thompson 1928, p. 30, pl. xxiii.24. (We question this assertion.)

28. At Armant (Mond and Myers 1937, pp. 255, 267, and 277.) a small fragment of a fin spine of Early Nagada II date was found in the settlement. At Hamamieh (Brunton and Caton-Thompson 1928, pp. 86, 94, 104) fish bones of *Lates* dating to early Nagada II were recovered on the floor of a hut. At Mostagedda, fish remains were found in the debris of village 300 (Brunton 1937, p. 58). Fish were also found in the settlements at Maadi along with copper fishhooks (Menghin and Amer 1932, p. 52).

29. McArdle 1982, pp. 116–121; Brewer 1987b, pp. 45–47.

30. Netolitzky 1911, pp. 953–956; 1943, pp. 8, 11–12.

31. Petrie 1921, pl. lii–lix; Baumgartel 1960, pp. 81–89.

32. Vandier 1952, pp. 382–384.

33. Bates 1917, pp. 204–206.

34. Hoffman 1982, cover; Červíček 1974, pp. 172–173.

35. Gamer-Wallert 1970, pp. 124–5.

36. Baumgartel 1960, p. 82. Gamer-Wallert 1970, pp. 25–27; Erman and Grapow 1926, pp. 264–267; Needler 1984, pp. 250–251. A slate *Tilapia* cosmetic holder from Abu Rowash dating to the First Dynasty contained a red cosmetic and no doubt has associations with the *Tilapia* in its breeding colors when it is known as the red fish.

37. Petrie 1921, pl. xviii; Vandier 1952, pp. 311–312; Bourriau 1981, pp. 30–33; see also Riefstahl (1972) for an alternate explanation.

38. Gamer-Wallert 1970, p. 64; Darby et al. 1977, pp. 379–380.

39. Gardiner 1976, pp. 476–477 (signs K 1–7); Montet 1914, pp. 39–48. See also Chapter III for an explanation of identifications.

40. Kaplony 1963, pp. 292–297, 993–994, and passim.

41. Petrie 1901b, p. 21, pl. 3a; Quibell and Green 1902, pl. 21.15.

42. Quibell and Green 1902, pl. xix.

43. See Chapter III under heading *Heterobranchus*.

44. Godron 1949, pp. 217–219; Kaplony 1963, ill. 5; Helck 1954, p. 969.

45. Gamer-Wallert 1970, p. 117; Altenmüller 1973, pp. 211–231.

46. Gamer-Wallert 1970, p.118; Emery 1938, p. 35. pl. 17–18, ill. 8.

47. Petrie 1901b, pl. vii. See also Chapter 2.

48. Emery 1962.

49. Gamer-Wallert 1970, pp. 66–67. (and see below)

50. Gaillard 1923, pp. 102–103, 125–126, and passim; Gamer-Wallert 1970, pp. 4, 15. See also Chapter III for interpretation.

51. Gaillard 1923, pp. 102–103, 111–112, and passim; Gamer-Wallert 1970, pp. 22–23. By the end of the Pharaonic period about 30 fish types are known and over 50 fish names of which 30 remain unknown.

52. Edel 1961, 1963; Smith 1965, pp. 141–147, pl. 178–179.

53. Darby et al. 1977, p. 362 – information gleened from the tombs of Ti and Mereruka. Macramallah 1935, pp. 42–43 – for the Tomb of Idout. The frequencies do not appear to be the same in the provinces where the *Lates* is depicted more commonly Davies 1902, II, p. 47.

54. Macramallah 1935, pp. 42–43, pl. 1.

55. – with the exception of those fish now extirpated or rare.

56. Darby et al. 1977, pp. 360–361; von Bissing 1905, pl. xvii.

57. Harpur 1987, pp. 148–149; Vandier 1969, pp. 635–666; Bates 1917, pp. 263–265; Davies 1902, pl. xx.

58. Other fish shown being prepared and dried include *Tilapia*, *Barbus* cf. *bynni*, *Lates*, *Citharinus*, *Mormyrus*, and *Synodontis*. The inclusion of *Synodontis* is interesting because structural differences of the skull would have made it difficult to prepare in the standard manner. (Dueli 1938, pl. 43; Ephron and Daumas 1939, pl. xi; Lepsius 1849, vol. II, pl. 9, pl. 46).

59. Although no depictions of the oblong shaped ovaries appear prior to the Dynasty V, earlier *Mugil* gutting scenes may have dipicted the row in paint (Keimer 1939; Harpur 1987).

60. Vandier (1964) believes scenes (Tombs of Nebemakhet at Giza and Urarna, Sheikh Said) that show roe being placed in a jar and then removed, represent pickling in brine. Scenes from Sheikh Said (fig. 1.8) and Hetep-her-akhty (Leiden) may show pressing and drying of the roe. See also van der Walle 1978, pp. 70–71, pl. 13.

61. Moussa and Altenmüller 1977, p. 100; Davies 1901, pl. xii; Hassan 1975, p. 36, fig. 16; Keimer 1939.

62. Montet 1925, pp. 40–41; von Bissing 1905, p. 11, pl. 17–19; Guglielmi 1973, pp. 161–162. A similar scene probably in Ti, (Ephron and Daumas 1939, pl. 111).

63. Hassan 1938, p. 520, ill.96; Smith 1942, pp. 315–316, fig. 5.

64. Moussa and Altenmüller 1977, pp. 81–83.

Janssen 1975, p. 348–350.

65. Simpson 1976, p. 2.

66. Moussa and Altenmüller 1977, pp. 59–60, ill. 5; Kanawati 1983, IV pp. 19–21, fig. 12; Simpson 1976, pp. 2–3, fig. 16; van de Walle 1978, pp. 65–70, fig. 12.

67. Harpur 1987, p. 114. In this context it is interesting to note that at the western Delta site of Kom el-Hisn, which may have been dedicated to the production of provisions for larger population centers and perhaps funerary estates, the inhabitants depended on sheep/goat, pig, and fish. Cattle, the predominant taxon recovered from tombs was remarkedly rare (Wenke 1986, pp. 15–33).

68. Bates 1917, p. 216.

69. For example, Moussa and Altenmüller 1977, pp. 90–96, fig. 12, depicting fish and fowl in parallel scenes being presented to the tomb owners with identical captions. See also Handoussa 1988.

70. Lichtheim 1975, pp. 163–169.

71. Gamer-Wallert 1970, p. 80.

72. Lacau 1913, pp. 42–49.

73. Montet 1950.

74. Edel 1961, 1963, 1976, 1978.

75. Boessneck and von den Driesch 1982, pp. 116–119. Although the journey of the *Mugil* to Elephantine was celebrated, few may actually have made the treck. The *fahaka* was also the subject of veneration at Elephantine, but its absence from the faunal record may be due more to its toxic properties than religious constrictions.

76. Blackman 1953, p. 32, pl. xxiv; Davies 1902, p. 13, pl. iv.

77. Guglielmi 1973, pp. 157–176; Bates 1917, pp. 268–269; Jaroš-Deckert 1984, p. 56ff, pl. 20.

78. Gamer-Wallert 1975a, p. 226.

79. Lichtheim 1975, p. 106; Boessneck and von den Driesch 1982; von den Driesch 1986, pp. 7–8.

80. Edel 1980; Jaroš-Deckert 1984, p. 32ff., pl. 16.

81. Caminos 1956, pp. 22–39.

82. Gamer-Wallert 1970, pp. 69–70.

83. Farag 1982.

84. Vandier 1969, pp. 601–605, 611–623.

85. Blackman 1953, pl. 11.

86. Numerous other interpretations are possible and have been offered. A religious association of fish with fertility and rebirth by extension included women and the fish shaped ornaments worn by women in this period may have been viewed in this context. Fischer 1977, pp. 164–165; Staehelin 1978.

87. Gamer-Wallert 1970, pp. 124–131; Vandier 1969, pp. 605–635; Desroches-Noblecourt 1954, pp. 33–42.

88. Gamer-Wallert 1970, pp. 65–66.

89. Gamer-Wallert 1970, pp. 71–72.

90. Gamer-Wallert 1970, p. 68; Darby et al 1977, pp. 382–383; Breasted 1906, Vol 4, pp 134–190.

91. Breasted 1906–7, vol. III, p. 207.

92. Christophe 1967, pp. 177–199.

93. Helck 1964, pp. 226–228; Gamer-Wallert 1970, pp. 24–46. A gutted *Alestes* was found an offering in a tomb of one of the workmen Keimer 1939, pl. 8–9; Keimer 1948.

94. Davies 1903, pl. 31.

95. Breasted 1906–7, vol. 4, p. 466.

96. Gamer-Wallert 1970, pp. 71–72.

97. von den Driesch 1983, table 2.

98. Janssen 1961, pp. 27–32, 96.

99. Darby et al. 1977, fig. 7.3 (Tomb Horemheb).

100. Janssen 1975, pp. 348–350.

101. Lichtheim 1976, p. 227.

102. Gamer-Wallert 1970, pp. 76–85.

103. Loat 1904.

104. Diodorus Siculus I, 36.

105. Herodotus II, ch. 149.

106. Herodotus II, ch. 77 and 92.

107. Gamer-Wallert 1975a, p. 226.

108. Darby et al. 1977, pp. 361–363.

109. Herodotus II, ch. 65; Diodorus Siculus I, ch. 83; etc.

110. van der Horst 1982, p. 66.

111. Herodotus II, ch. 37.

112. Lichtheim 1980, p. 80.

113. Gamer-Wallert 1970, pp. 75–85.

114. Montet 1950. Gamer-Wallert 1970, pp 81–85.

115. Plutarch, Isis and Osiris. 355, 1 D–358, 19E.

116. Lichtheim 1976, pp. 203–211.

117. Altenmüller 1973.

118. Aelian X, ch. 36.

119. Gamer-Wallert 1970, pp. 91–95.

120. Ibid.

121. Gamer-Wallert 1970, pp. 95–98.

122. Aelian X, 19; Plutarch Isis and Osiris 358, 18b.

123. Edel 1961, 1963, 1976.

124. Gamer-Wallert 1970, pp. 101–107.

125. Rawlinson 1866, vol. II, p. 104, 107. Gamer-Wallert 1970, pp. 115–116.

126. Lloyd 1982.

127. Keimer 1948.

128. Gamer-Wallert 1970, pp. 88–9 109–114; Desroches-Noblecourt 1966; Staehelin 1978.

129. Gamer-Wallert 1970, pp. 82–8 Gamer-Wallert 1975b, p. 233.

130. Hornung 1982, pp. 181–182.

131. Hornung 1982, pp. 100–101, 137–13 Kessler 1985, pp. 571–587.

132. Gamer-Wallert 1975b, p. 229.

133. Lagercrantz 1953, pp. 1–8.

CHAPTER II

1. The first undisputed evidence for the rod comes from the 12th dynasty tomb of Khnumhotep, Beni Hasan. Newberry and Fraser 1893, pl. XXIX. See also fig. 2.26 and 46ff. this chapter.

2. The implements that have been recovered are the by-products of other archaeological research interests and may not be representative of actual ancient Egyptian practices.

3. Harpoon points of bone, horn, and ivory are "common" at Nagada only in terms of their relative frequency vis-à-vis other sites. At Nagada 10 harpoon heads were found in one grave (N1215), two others were found in settlement debris (South-town). Examples of harpoon heads from other sites include two from El-Amrah, one from Mahasna, one from Ballas, and one from El-Adaima. Petrie and Quibell 1896, p. 46; Petrie 1920, p. 24; Needler 1984, p. 289–90.

4. Merimde has yielded bone points from all levels. Points recovered from the lowest level, however, could have been employed as awls. Barbed harpoon heads (three) first appear in the second level of the recent excavations. Junker 1929, pp. 156–250; Larsen 1960, 28 ff; Eiwanger 1979, p. 53; Eiwanger 1984, pp. 55–57; Debono and Mortensen in press.

5. Bates 1917, p. 233.

6. Vandier 1952, figs. 184, 189, 190.

7. Bates 1917, p. 237.

8. Petrie 1915, p. 15.

9. Petrie 1920, p. 24.

10. Petrie 1901a; Petrie 1920, p. 24.

11. Petrie 1917, p. 37.

12. Bates 1917, p.240.

13. The general form of the Pre- and Protodynastic harpoons clearly survived in copper until at least the 18th Dynasty (e.g., fig. 2.2).

14. Petrie 1917, p. 37; Petrie 1901b, pls. XXXV.92, XLIV.12, and pl IX.a.

15. Baumgartel 1960, p. 18.

16. The retrieving line becomes more common in the Middle Kingdom. Säve-Soderbergh 1953, pp. 1012–1015. For provincial variants with retrieving line see Vandier 1969, p. 730.

17. Säve-Soderbergh 1953, pp. 6–23.

18. ... But see fish spearing scene in Tomb of Henku, Deir el-Gebrâwi were attendant is shown holding a large single barb harpoon for the deceased. Davies 1902, pl. XXIII.

19. e.g., The tomb of Sebeknakht, El-Kab. Tylor 1896, pl. III.

20. Säve-Soderbergh 1953, pp. 6–25.

21. Gardiner 1976, p. 514; Fischer 1983, p. 42.

22. Bates 1917, p. 243.

23. It is possible that a reference to the use of the bident may have been present in the bibliographic inscription of Henku. Unfortunately, the text is not well preserved. See Bates 1917 p. 244, 167ff.

24. Petrie 1902, pl. XI; Petrie 1901b, pl. VII.

25. Daumas 1975, p. 235; Vandier 1964 pp. 717–772 – especially 731 f. and references cited there and pl. 399.2; Borchardt 1913, pl. XVI.

26. See Gamer-Wallert 1970, p. 73f, 129f for a religious interpretation of scenes for all periods. Desroches-Noblecourt 1954, pp. 33–42.

27. Examples that deviate from the common *Tilapia-Tilapia* or *Tilapia-Lates* combination include, but are not limited to: a single speared fish (Tilapia) from the tomb of Khety, Beni Hassan and a fish other than the *Tilapia* or *Lates* shown in the tomb of Khezou-wer, Kom el-Hisn (*Tilapia* and unidentified fish) (Vandier 1964, fig. 400.1). See Moussa and Altenmüller 1977, pl. 4, scene 4.3.1 for the only Old Kingdom example showing two speared Tilapia (The Tomb of Two Brothers).

28. Kanawati 1980–5 (vol. II) p. 26. fig. 22. Vandier 1964, p. 729f. and fig. 411. Duell 1938, vol II, pls 127, 128. See also fig. 2.6 this chapter. For a list of Old Kingdom harpooning scenes see Harpur 1987, pp. 355–367, table 7; for the Middle Kingdom see Vandier 1964 pp. 758–759; and for the New Kingdom see Vandier 1964, pp. 758–759 and Porter and Moss 1960, table 17a.

29. Bates 1917, p. 244.

30. i.e., related to other sites. Krzyzaniak 1977, pp. 71, 104, 142. Brunton 1937, p. 56.

31. Caton-Thompson and Gardner 1934, pp. iii–167.

32. Wenke and Lane (in press)

33. Junker 1929, pp. 156–250; Eiwanger 1979, p. 54; Eiwanger 1984, pp. 56–58.

34. Debono 1948, pp. 561–559; Debono and Mortensen (in press), pl. 26.40.

35. de Morgan in Radcliff 1926, p. 86; Rizkana and Seeher 1988, pl. 9, 25, 31–36.

36. Personal observation, see also Wenke and Lane (in press).

37. Needler 1984 and references cited there. Baumgartel 1960, pp. 18–20.

38. Kromer 1978, p. 81, pl. 32.6–7.

39. Bates 1917, p. 246 and figures referenced there; see also Petrie 1917, p. 37 and plates referenced there. Touny and Wenig 1969, pp. 67–68

40. Bates 1917, p. 247.

41. Petrie 1890, p. 34.

42. Bates 1917, p. 248.

43. Approximately 20 Old and Middle Kingdom tombs contain angling scenes. Moussa and Altenmüller 1977, p. 97, note 485, fig. 12. Vandier 1969, pp. 533–547; Kanawati 1980, I fig. 12; II fig. 22; IV fig. 20–21.

44. It is interesting to note that *Synodontis* is easily caught by hook and line. Boulenger 1907, p. 357; Brewer personal observation.

45. Upon hooking a *Tetraodon* the unfortunate angler in the Tomb of Djaou at Deir el Gebrâwi exclaims "Is this fish?" (Davies 1902, II, pl. IV) For others see Daumas 1964 and Moussa and Altenmüller 1977.

46. Although it may appear a piece of local color, the catch is shown to be expressly for the purpose of supporting the mortuary estate or sustaining eternal nourishment for the tomb owner (von Bissing 1905, pl. 1).

47. Vandier 1969, pp. 533–541.

48. Contrary to Harpur (1987) in the Tomb of Ankhtify (Dyn. IX), the tomb owners wife is not engaged in fishing, but rather, is holding a duck. See also Vandier 1950, pl. XIII.

49. Five such scenes are known. (Vandier 1969, pp. 605–609).

50. Gamer-Wallert 1970; Desroches-Noblecourt 1954, pp. 33–42.

51. Radcliff 1926, p. 311 suggests it as a possibility.

52. In the Tomb of Two Brothers a club can be seen laying near the fisherman's feet, suggesting it was not employed as a reel (Moussa and Altenmüller 1977, pp. 98, pl. 12.).

53. A reel, similar to those used for hippo hunting is shown being employed in spear fishing in the Tomb of Ankhtifi (Vandier 1950, pl. X).

54. Wilkinson 1883 (vol. II), p. 116.

55. In several scenes (e.g., fig. 2.9) an object is shown attached near the gang hook. It is possible that these objects, usually identified as weights or sinkers, might actually represent bags containing bait. See Daumas 1964.

56. Budge 1967, p. 346 (text cxxv, intro).

57. Caminos 1956, pp. 1–21.

58. Brewer 1986, pp. 152–169 and references cited therein.

59. Ruffer 1919, p. 32.

60. Winkler 1938 (vol. I) p. 31.

61. Strabo 1932, p. 153.

62. Loat 1907, p. xxxi.

63. Meeks 1973, pp. 209–216.

64. Darby et al. 1977, p. 341.

65. Loat 1907, p. xxxvii.

66. Vandier 1969, pp. 551–559. Moussa and Altenmüller 1977, pp. 98–100.

67. A weir might be portrayed in the Middle Kingdom tomb of Djehutyhotep (tomb 2) at El-Bersheh, but the scene is too damaged for a positive identification. Newberry 1893 (I), p. 9.

Weirs (nassa) may have been represented on pottery vessels of the Predynastic. The depictions are, however, quite stylized and fish are never associated with the figures. Consequently, a positive identification is not possible. Scamuzzi 1964, pl. IV; Petrie 1920, pl. 23;

68. Edel 1963, p. 208, fig. 18.

69. Associated with the men using the small weirs in the tomb of Ti are recorded fragments of their conversation. In one instance a man about to lift a weir out of the water says to his companion bringing a basket, "Are you, you jerk, trying to tell me my own business?" The response is; "Drop it and hurry up." Guglielmi 1973, p. 159. .

70. Moussa and Altenmüller 1977, pp. 98–99 and references cited there; Guglielmi 1973, p. 159; Ephron and Daumas 1939, pl. III.

71. Lacau 1954, pp. 136–163; Vandier 1969, pp. 547–551, 609–611.

72. The third example of the basket trap is in the Dynasty XI tomb of Djaou, Thebes. Basket traps were also used to transport the catch. See Vandier 1969 for details.

73. Caminos 1956, pp. 1–21; Lacau 1954, p. 140.

74. ... and worked in a variety of ways. See Vandier 1969, pp. 559–635 for details on specific examples of the net in use and the juxtaposition of the fishermen involved.

75. Debono and Mortensen (grave A89) (in press); Mortensen, pers. comm.

76. Radcliff 1926, p. 316, 3ff.; Petrie 1892.

77. Petrie 1890, p. 34.

78. We are aware of 13 Old Kingdom tombs and 4 Middle Kingdom tombs depicting the handnet. No handnet scenes are known from the New Kingdom. Vandier 1969, 541ff; Moussa and Altenmüller 1977, p. 98, pl 12; Kanawati 1980–, I, pp. 22–25, fig 12; II, pp. 26–28, fig. 22.

79. Dunham 1935, pp. 300–309.

80. Brewer personal observation. See also Henein 1988, pp. 133–135, 137, fig. 130. Loat 1907, pp. XXXIX–XL, fig. 23.

81. Loat 1907, pp. xxi–xxii.

82. See section on seines, this chapter, for further clarification.

83. Bates 1917, p. 257.

84. Seines are represented in at least 78 Old Kingdom tombs. Harpur 1987, table 6.20; New Kingdom, Porter and Moss 1960, table 17e; Vandier 1969, pp. 613–635.

85. Net weights are frequently recovered from Prehistoric sites. Net weights of limestone, grooved along the middle, were relatively abundant at el-Omari. At Merimde net weights made of stone and ceramic were recovered from most levels. Net weights have also been recovered from the Fayum, and Khartoum. Debono and

Mortensen (in press), pl. 32. 4–7; Eiwanger 1984, p. 58; Caton-Thompson and Gardner 1934, pp. 39–40, pl. XXIX.7–9; Arkell 1949, pl 26.17–20, 40.3–4.

86. For ancient respresentations of playing out (or gathering in) the net from boats see depictions from the sun temple of Niussere (Smith 1965, fig. 178) and from the Tomb of Huya at Amarna (Davies 1905, pl. 8).

87. Harpur 1987, pp. 145–148, 175–221.

88. Bidoli 1976, pp. 473–480.

CHAPTER III

1. The marine fish represented on the temple of Hat-shepsut, Deir el-Bahri, are not considered in this volume. Interested readers are referred to Danelius and Steinitz (1967, pp. 15–24) for identifications of these taxa.

2. Sandon 1950, p. 21; Latif 1974, p. 19; Greenwood 1966, p. 21.

3. Greenwood 1966, pp. 21–22; Hopson 1982, p. 292.

4. Greenwood 1966, pp. 21–22.

5. Greenwood 1966, pp. 21–22.

6. Greenwood 1966, p. 22.

7. Boulenger 1907, p. 35; Latif 1974, p. 22.

8. Latif 1974, pp. 28–29; Greenwood 1966, pp. 22–23; Boulenger 1907, pp. 32–39.

9. Sandon 1950, pp. 21–22; Abu-Gideiri 1984, pp. 25–26.

10. Sandon 1950, pp. 21–22.

11. Boulenger 1907, p. 39.

12. Boulenger 1907, p. 35.

13. Latif 1974, p. 21.

14. Boulenger 1907, pp. 50–53; Latif 1974, p. 24.

15. Latif 1974, p. 25.

16. Boulenger 1907, p. 52.

17. Sandon 1950, pp. 23–24.

18. Boulenger 1907, p. 50; Greenwood 1966, p. 34; Latif 1974, p. 24; Abu-Gideiri 1984, p. 28.

19. Boulenger 1907, p. 61; Sandon 1950, p. 30.

20. Reported by Boulenger 1907, p. 62.

21. Corbet 1961; Greenwood 1966, pp. 25–28.

22. Latif 1974, pp. 26–28.

23. Greenwood 1966, pp. 25–28.

24. Latif 1974, pp. 27–28.

25. Abu-Gideiri 1984, pp. 26–28.

26. Boulenger 1907, p. 59.

27. Boulenger 1907, pp. 61–69; Greenwood 1966, pp. 25–28; Latif 1974, pp. 25–28.

28. Greenwood 1966, p. 23; See also Greenwood 1963, pp. 61–74.

29. Daget 1954.

30. Sandon and Tayib 1953, p. 211.

31. Boulenger 1907, p. 73.

32. Latif 1974, p. 29.

33. Boulenger 1907, pp. 70–73; Greenwood 1966, pp. 22–23; Latif 1974, p. 29.

34. Boulenger 1907, pp. 98–109; Greenwood 1966, pp. 35–38; Latif 1974, pp. 31–35; Abu-Gideiri 1984, pp. 51–53.

35. Boulenger 1907, pp. 104–106; Latif 1974, pp. 33–34.

36. Sandon 1950, p. 53.

37. Boulenger 1907, pp. 98–109.

38. Greenwood 1966, pp. 36–38.

39. Boulenger 1907, p. 101.

40. Greenwood 1966, p. 36.

41. Latif 1974, pp. 33–35.

42. re: Chapter 1

43. Keimer 1948, pp. 263–274.

44. Sandon 1950, p. 28; Abu-Gideiri 1984, p. 54;

45. Latif 1974, p. 31.

46. Boulenger 1907, p. 110.

47. Boulenger 1907, p. 130.

48. Latif 1974, p. 39.

49. Boulenger 1907, pp. 110–130; Sandon 1950, pp. 54–55.

50. Latif 1974, p. 39.

51. Sandon 1950, p. 56.

52. Boulenger 1907, pp. 152–153.

53. Boulenger 1907, p. 153.

54. Boulenger 1907, pp. 153–158; Latif 1974, pp. 42–44.

55. Latif 1974, pp. 42–44; Bishai Personal Communication 1987.

56. Latif 1974, pp. 42–44; Boulenger 1907, pp. 153–158.

57. Greenwood 1966, p. 46.

58. Daget 1954; Stevenson 1933 in Greenwood 1966, p. 46.

59. Boulenger 1907, p. 158.

60. Abu-Gideiri 1984, pp. 63–64.

61. Sandon 1950, p. 34.

62. Latif 1974, pp. 50–51.

63. Boulenger 1907, p. 161; Greenwood 1974, p. 50;

64. Latif 1974, p. 50.

65. Greenwood 1966, p. 50–53; Sandon 1950, p. 36.

66. Boulenger 1907, p. 166; Hashem 1973, p. 85.

67. Hashem 1973, p.85.

68. Sandon 1950, p. 35.

69. Boulenger 1907, p. 161.

70. Greenwood 1966, pp. 50–53; Boulenger 1907, p. 161 Latif 1974, pp. 52–56; Sandon and Tayib 1953, pp. 216–217.

71. Reported by Boulenger 1907, p. 166.

72. Latif 1974, pp. 52–56.

73. Gaillard 1923, p. 40.

74. Latif 1974, p. 58.

105

75. Greenwood 1966, p. 58.

76. Greenwood 1966, p. 58; Boulenger 1907, pp. 198–199.

77. Greenwood 1966, pp. 58–73.

78. Latif 1974, p. 58.

79. Sandon and Tayib 1953, p. 217.

80. Boulenger 1907, pp. 195–260.

81. Sandon 1950, p. 36.

82. Boulenger 1907, pp. 195–260; Latif 1974, pp. 58–62.

83. Latif 1974, pp. 58–63; Boulenger 1907, pp. 195–260.

84. Latif 1974, p. 59.

85. Hashem and El-Agamy 1977, p. 145; see also pp. 145–149.

86. Elster and Jensen 1960, p. 91.

87. Greenwood 1966, p. 63; Pekkola 1919, p. 116.

88. Loat 1907, p. 204.

89. Boulenger 1907, pp. 278–299.

90. Greenwood 1966, p. 81

91. cf. Bruton 1979, pp. 7–9.

92. Greenwood 1966, p. 81; Greenwood 1968, pp. 100–109; see also Roberts 1975, pp. 249–319.

93. Greenwood 1966, pp. 81–82.

94. Greenwood 1966, p. 82.

95. Reported in Boulenger 1907, p. 281.

96. Reported in Sandon 1950, p. 41.

97. Boulenger 1907, p. 279; Latif 1974, p. 65.

98. Latif 1974, p. 65.

99. Boulenger 1907, pp. 280 and 288.

100. Personal interviews 1983, 1987; see also Pekkola 1919, p. 117; Boulenger 1907, p. 281.

101. Pekkola 1919, p. 117.

102. Sandon and Tayib 1953, pp. 217–218.

103. Brewer 1986, p. 83.

104. Sandon and Tayib 1953, p. 217.

105. Greenwood 1966, p. 85.

106. Reported in Greenwood 1966, p. 85.

107. Boulenger 1907, p. 281; Greenwood and Todd 1970.

108. Brewer 1986, p. 84.

109. Greenwood 1966, p. 84; Greenwood 1955, pp. 516–518.

110. Bruton 1979, pp. 19–21.

111. Bruton 1979, pp. 1–45, esp. 19–21; Whitehead 1959, pp. 329–363; Qasim and Qayyam 1961, pp. 24–43.

112. Sandon 1950, pp. 41–42; Latif 1974, pp. 68–70.

113. Boulenger 1907, p. 300.

114. Reported in Greenwood 1966, p. 87.

115. Quibell and Green 1902; Quibell 1900.

116. Sandon 1950, pp. 49–50; Latif 1974, pp. 70–77; Boulenger 1907, pp. 311–317.

117. Greenwood 1966, pp. 78–79; Corbet 1961; Latif 1974, pp. 73–74.

118. Latif 1974, p. 75.

119. If any representations of *Schilbe* possess an adipose fin, they would be placed under the genus *Siluranodon*.

120. Latif 1974, p. 77; Sandon 1950, pp. 42–43.

121. Sandon 1950, p. 77; Boulenger 1907, p. 323.

122. Reported in Boulenger 1907, pp. 327 and 330.

123. Corbet 1961; Greenwood 1966, pp. 175–177.

124. Sandon and Tayib 1953, p. 218.

125. Brewer, personal observations and interviews.

126. Boulenger 1907, p. 327.

127. Latif 1974, pp. 78–79; Sandon 1950, p. 43.

128. Latif 1974, pp. 79–80.

129. Boulenger 1907, pp. 323–333.

130. Pekkola 1919, p. 117.

131. Latif 1974, p. 79.

132. Sandon 1950, p. 44.

133. Boulenger 1907, pp. 351–352.

134. Bishai and Abu-Gideiri 1967, p. 20, Fig. 3.

135. Greenwood 1966, p. 88.

136. Latif 1974, p. 87.

137. Based on experience, the flavor of the fish is less than palatable to Western European tastes. Brewer, personal observation 1984.

138. Brewer 1986, p 86.

139. Boulenger 1907, p. 351.

140. Bishai and Abu-Gideiri 1965, pp. 98–112.

141. Pekkola 1919, pp. 117–118; Sandon and Tayib 1953, pp. 219–221.

142. Reported in Boulenger 1907, p. 379.

143. Reported in Boulenger 1907, p. 382.

144. Bishai and Abu Gideiri 1965, pp. 85–113.

145. Sandon 1950, p. 45.

146. Sandon 1950, p. 45; Bishai and Abu Gideiri 1965, pp. 85–113.

147. Luxor Fish Market, Brewer 1986. The ancient Egyptian medical texts mention the use of various parts of the *Synodontis* in remedies; the skull was used to treat headaches, the brains (?) for maladies of the hair and the blood was used in eyelash treatments (Gammer-Wallert 1970, p. 69).

148. Brewer, personal interview 1987; Latif 1974, p. 94.

149. Latif 1974, p. 94; Boulenger 1907, p. 395; Greenwood 1966, pp. 93–94.

150. Boulenger 1907, p. 394.

151. Greenwood 1966, p. 94.

152. Sandon 1950, p. 50.

153. Boulenger 1907, pp. 399–400.

154. Greenwood 1966, p. 94.

155. Latif 1974, p. 95.

156. Brewer, personal interview 1986, 1987

157. Boulenger 1907, p. 402.

158. Boulenger 1907, p. 404.

159. Boulenger 1907, p. 404.

160. Reported in Boulenger 1907, p. 405.

161. Boulenger 1907, p. 427.

162. Reported in Boulenger 1907, p. 428.

163. Boulenger 1907, p. 434.

164. Reported in Boulenger 1907, p. 436.

165. Boulenger 1907, p. 428. One of the ancient Egyptian names for the mullet means 'jumper' (Edel 1980).

166. Brewer, personal interview 1986. Edel (1963, pp. 160–161) states that mullets swim upriver from Lake Manzala at the end of May and return to the sea to spawn by late November. This is slightly contradictory to Boulenger (1907, p. 428) who noted that mullet fry entered the lake in June. It may be that mullet enter the lakes and river areas from January through June.

167. Boulenger 1907, p. 428.

168. Gaillard 1923, pp. 90–96.

169. Boulenger 1907, pp. 427–428.

170. Sandon 1950, p. 53.

171. Boulenger 1907, p. 452; Latif 1974, p. 102.

172. Sandon 1950, p. 53.

173. Latif 1974, p. 103 and Fig. 60.

174. Latif 1974, p. 103.

175. Greenwood 1966, pp. 103–105; Roberts 1975, p. 260

176. Fish 1956, pp. 186–190.

177. Reported in Boulenger 1907, p. 455.

178. Hashem and Hussein 1973, p. 389.

179. Latif and El-Sayed 1974, p. 103.

180. Kenchington 1939, p. 160.

181. Hopson 1972.

182. Pekkola 1919, p. 119.

183. Reported in Sandon and Tayib 1953, p. 223.

184. Hopson 1972; See also Latif and Khallaf 1974, p. 144.

185. Latif and Khallaf 1974, p. 144, pp. 154–155; Hashem and Hussein 1973, p. 391.

186. Latif and Khallaf 1974, p. 139.

187. Hashem and Hussein 1973, p 391.

188. Latif 1974, p. 95; Boulenger 1907, p. 459.

189. Latif 1974, p. 95.

190. Greenwood 1951, pp. 19–20; Greenwood 1956, pp. 233–244; Greenwood 1965, pp. 256–269; Greenwood 1966, pp. 106–107. Brooks 1950, p. 131–176.

191. Sandon 1950, p. 54.

192. Greenwood 1966, p. 106; Latif 1974, p. 96.

193. Greenwood 1966, p. 106; Latif 1974, p. 108; Elster and Jensen 1960, p. 63.

194. El-Zarka et al. 1970, p. 151.

195. Latif 1974, p. 96.

196. Boulenger 1907, pp. 513–514.

197. Greenwood 1966, p. 110; Brewer, personal observation 1986.

198. Sandon 1950, p. 56.

199. Sandon 1950, p. 56.

200. Latif 1974, pp. 97–103; Sandon and Tayib 1953, pp. 223–224.

201. Quibell 1900, pl xxii.

202. e.g., Tomb of Nakht, Dyn XVIII, Sheikh abd el Qurna, T–52; Tomb of Khnumhotep Dyn. XII, Beni Hasan; Tomb of Pepi Honk Henit Dyn VI, Meir; Tomb of Mereruka Dyn VI, Saqqara.

203. Latif 1974, p. 98.

204. Sandon and Tayib 1953, pp. 223–224.

205. Sandon and Tayib 1953, pp. 223–224; Greenwood 1966, pp. 109 and 111; See also Fish 1955, p. 189.

206. Sandon and Tayib 1953, pp. 223–224; Greenwood 1966, p. 109.

207. Latif 1974, p. 98.

208. Latif 1974, p. 99.

209. Sandon and Tayib 1953 , pp. 223–224; Greenwood 1966, p. 109; Latif 1975, p. 98.

210. Boulenger 1907, p. 523.

211. Brewer, personal observation 1984, 1986, 1987, 1988.

212. Sandon 1950, p. 60.

213. Boulenger 1907, pp. 544 and 546.

214. Latif 1974, p. 107.

215. Latif 1974, p. 107; Pekkola 1919, p. 120; Sandon and Tayib 1953, p. 225.

216. Bishai, personal communication 1987.

217. Latif 1974, p. 105.

218. Sandon 1950, p. 60.